Chemistry Laboratory

for Secondary and Higher Education

3rd Edition

Toshihiko SONOBE and Fumio KAWAIZUMI

園部 利彦　　　川泉 文男

Gakujutsu Tosho Shuppan-sha Co. Ltd.
学術図書出版社

About the Authors

Toshihiko SONOBE taught chemistry for 36 years at public high schools in Gifu, Japan. He was given The Chemical Society of Japan Award for Merit of Chemical Education for 2003.

Fumio KAWAIZUMI was a professor at Graduate School of Engineering, Nagoya University, Japan. His major is physical chemistry and fundamentals of chemical engineering. He earned The Chemical Society of Japan Award for Chemical Education for 2001.

◇

The authors express their deep gratitude to Mr. Paul A. CRANE, an instructor of English living in Japan for almost ten years and working for several Nagoya area universities, for his helpful advice and editorial supervision on English of the 1st edition of this book.

Contents

Periodic Table

	1	2	3	4	5	6	7	8	9	10	11	12	13	14	15	16	17	18
1	1H		Nonmetallic-representative elements				Metallic-representative elements											2He
2	3Li	4Be	Metallic-transition elements				State-unknown elements						5B	6C	7N	8O	9F	10Ne
3	11Na	12Mg											13Al	14Si	15P	16S	17Cl	18Ar
4	19K	20Ca	21Sc	22Ti	23V	24Cr	25Mn	26Fe	27Co	28Ni	29Cu	30Zn	31Ga	32Ge	33As	34Se	35Br	36Kr
5	37Rb	38Sr	39Y	40Zr	41Nb	42Mo	43Tc	44Ru	45Rh	46Pd	47Ag	48Cd	49In	50Sn	51Sb	52Te	53I	54Xe
6	55Cs	56Ba	57-71	72Hf	73Ta	74W	75Re	76Os	77Ir	78Pt	79Au	80Hg	81Tl	82Pb	83Bi	84Po	85At	86Rn
7	87Fr	88Ra	89-103	104Rf	105Db	106Sg	107Bh	108Hs	109Mt	110Ds	111Rg	112Cn	113Nh	114Fl	115Mc	116Lv	117Ts	118Og

57-71 Lanthanoid	57La	58Ce	59Pr	60Nd	61Pm	62Sm	63Eu	64Gd	65Tb	66Dy	67Ho	68Er	69Tm	70Yb	71Lu
89-103 Actinoid	89Ac	90Th	91Pa	92U	93Np	94Pu	95Am	96Cm	97Bk	98Cf	99Es	100Fm	101Md	102No	103Lr

As for the history of elements, see the description given on the page inserted between pages 10 and 11.

Name of Element	Symbol	Approximate Atomic Weight	Name of Element	Symbol	Approximate Atomic Weight
Aluminum	Al	27.0	Lead	Pb	207
Barium	Ba	137	Lithium	Li	6.9
Bromine	Br	79.9	Magnesium	Mg	24.3
Calcium	Ca	40.1	Manganese	Mn	54.9
Carbon	C	12.0	Mercury	Hg	201
Chlorine	Cl	35.5	Nitrogen	N	14.0
Chromium	Cr	52.0	Oxygen	O	16.0
Cobalt	Co	58.9	Phosphorus	P	31.0
Copper	Cu	63.5	Potassium	K	39.1
Fluorine	F	19.0	Silver	Ag	108
Hydrogen	H	1.0	Sodium	Na	23.0
Iodine	I	127	Sulfur	S	32.1
Iron	Fe	55.8	Zinc	Zn	65.4

Preface to This Edition

Chemistry is one of the bases of science. It plays an essential role in giving us information on the structure of matter on the microscopic level, properties and their relation to structure, and change of matter (reaction). Chemistry has created new and useful products for everyday life and has contributed to protecting the environment.

In learning chemistry, understanding what is written in textbooks and what is practiced in the laboratory is intimately related. Unfortunately, laboratory practice in chemistry is difficult to introduce into the curriculum of developing countries due to the lack of laboratory facilities and equipment, chemicals, and appropriate textbooks. One of the authors, F. K., who was involved in improving chemistry education in Cambodia, realized the difficulties mentioned above.

This book was prepared with the dual intention to stimulate young students, including those in developing countries; to get skilled in laboratory practices and to learn English through chemistry. The first version of this book was "Chemistry Laboratory for Secondary and Higher Education in Developing Countries" published in 2002. It was transformed to New Edition (1st edition of this book) in 2004 and then modified in 2006. The descriptions and contents have been drastically modified in each revising process to match our intention of publication. Color photographs are included in this edition.

We are now in the 21st century, in which various environmental problems will strongly menace the welfare and future of our lives. There are global ones such as the depletion of the ozone layer and local ones such as the treatment of waste water and garbage created by urban areas. In the 21st century, deep understanding of science and its relation to human life is far more important than it was in the 20th century. This is especially the case in developing countries. As authors living in Japan and passengers of "Space Ship Earth", we would be happy if this book could contribute to the education of chemistry and the result is the peace and happiness of our fellow creatures.

January 2010   Toshihiko SONOBE and Fumio KAWAIZUMI

Instructions for Laboratory Practices

1 Before starting experiments

1) Read through the description in this book carefully so as to understand the purpose and method of the experiment. "Know why" is more essential than "know how" in manual-reading.

2) Follow the instructions of your teachers. Unless you are instructed differently, never try to do what is not described in this book.

3) In the case of an accident, even if it is small, never fail to inform your teachers of it.

2 Precautions in experiments

1) To ensure your own safety, wear protective clothing and safety goggles at all times. When handling concentrated acids, you should wear gloves.

2) Observe quietly and seriously. Always keep in mind the purpose of the experiment while paying attention to the operation. Record the results of the experiment without fail. Be sure what you have recorded is thorough and complete.

3) Follow the instructions of your teacher in discarding the waste. Do not discard solid waste into the sink.

4) Do not use an excess amount of reagents in order to prevent pollution by their waste.

5) Use separate spatula for each chemical when you take out several kinds of chemicals at a time. Once you take out a reagent from the bottle, do not return it to the bottle.

3 After experiments

1) Wash glassware and put it in the proper place. Discard dusts into a wastebasket. Wipe the work bench with a rag.

2) Check and examine the results you have obtained. Discuss them with your coworkers.

4 Reporting the results of experiments

1) Summarize the results by yourself. Refer to the textbook and reference books if necessary.

2) Write concisely. Keep the announced deadline of the report without fail.

3) Make an effort to find out the correct and proper answers to the questions.

Precaution against Accidents and First-aid Treatment

In chemistry experiments, you use many dangerous and/or harmful reagents. Accidents may occur even if you obey the instruction of your teacher and carry out the experiment attentively. Use the proper amount of chemicals as indicated in this manual. Failure to do so may cause the experiment to become inaccurate and dangerous. When an accident occurs, keep a cool head.

Accidents that may occur in the chemistry laboratory

1 Bumping : Bumping is the phenomenon in which the contents in a test tube, beaker, or flask boil suddenly when you are heating them. When bumping occurs, the heated liquid boils over. At times it might jump out several meters. The liquid that jumps out may burn you or ignite. While heating the liquid in a test tube you must not point the test tube at someone else. To avoid bumping, put one or two pieces of porous boiling stone in it before heating and heat the test tube while shaking it to swirl the liquid.

2 Fire : Keep flammable liquid (such as hydrocarbon, alcohol, ether, ester etc.) away from naked fire because the vapor or liquid may ignite. Do not put any combustibles on a work bench during an experiment. Do not use a burner in the place where strong wind is blowing. Cover the container with appropriate incombustibles when liquids are burning in it. In the case where a small amount of liquid is burning, keep all flammables away from the fire and wait for it to burn out. Put out the fire with water, sand, or with a fire extinguisher.

3 Back-suction of water : Back-suction of water occurs as a result of reduced pressure in the container which is connected through a conduit to water. Therefore, the back-suction may happen when heating of the container, set in the aforementioned condition, is stopped and the container is cooled rapidly while one end of the conduit is dipped in water. Hence, before you stop heating the test tube or flask, you must take the conduit out of water.

4 Explosion : In handling hydrogen, fire should be kept sufficiently away because the inflammability limit is very wide (4 - 75%). Do not ignite hydrogen at the mouth of the conduit. You must collect hydrogen in a test tube before igniting. Alkali metals such as sodium and potassium must be used in small pieces in experiments. Do not dump their residue into the sink.

First-aid treatment in accidents

1 Burn : IMMEDIATELY LET COLD WATER (10 - 15 ℃) RUN OVER THE BURNT AREA FOR MORE THAN SEVERAL MINUTES. Put out burning clothes first. Cut or remove the clothes so as not to scar the burnt area. After cooling, wrap a few towel sheets moistened with water and ice, around the burnt area. Do not apply oil or tincture to the burnt area because they may provoke infection. When the victim is in shock, or is burnt extensively over his/her body, immediately call for a doctor to transfer him/her to the hospital.

2 Strong acids and alkalis

1) In the case where strong acids splash on the skin and clothes : Wash the affected part well with water for about 15 minutes. Then wash with a dilute solution of alkaline soap or the solution of sodium hydrogencarbonate. Do not neutralize the affected part right away without the above treatment, for the damage may be deteriorated by heat of neutralization. In the case where acids get into the eye, open the eyelid and wash with plenty of water for 15 minutes. Get to a physician as soon as possible. In the case where clothes are spattered, wash with water and then with dilute ammonia water.

2) In the case where strong alkalis splash on the skin and clothes : ALKALIS ARE FAR MORE HARMFUL TO EYES THAN ACIDS. In the case where alkalis get into the eye, follow the treatment described in **1)** above. In the case where clothes are spattered, undress immediately. Wash well with running water until the greasy feelings on the skin disappear. Next, neutralize with acetic acid or lemon juice diluted with water. In the case of splashing with quick lime (CaO, calcium oxide), brush the lime powder off of the clothing and the body. Then, wash with a plenty of water.

3) In the case where strong acids or strong alkalis are swallowed : TAKE THE VICTIM TO A DOCTOR WITHOUT DELAY.

3 Other chemicals : If the victim has a convulsive fit or loses consciousness, call for a doctor as soon as possible. Inform the doctor of the kind of chemicals, quantity, situation of accident, occurrence time, and symptoms. In the case of swallowing chemicals, take the means that lower the concentration in the stomach and delay absorption. In the case of inhaling harmful or stifling gases, immediately transfer the victim to the place where he/she can breathe fresh air. Perform artificial respiration on the victim. In the cases where chemicals damage clothes or skin, first-aid treatment is essentially the same as for accidents in **2** above.

Part I Essentials of Chemistry Laboratory

1 Name of Laboratory Ware

Test tube and test tube rack   Branched test tube   Test tube clamp   Spoon (Spatula)   Washing brush

Flask (flat bottom)   Flask (round bottom)   Flask (branched)   Erlenmeyer flask   Conical beaker   Beaker

Funnel and funnel stand   Separatory funnel   Büchner funnel and suction flask   Volumetric flask (Measuring flask)   Bottle

Petri dish   Watch glass   Dropping bottle   Washing bottle

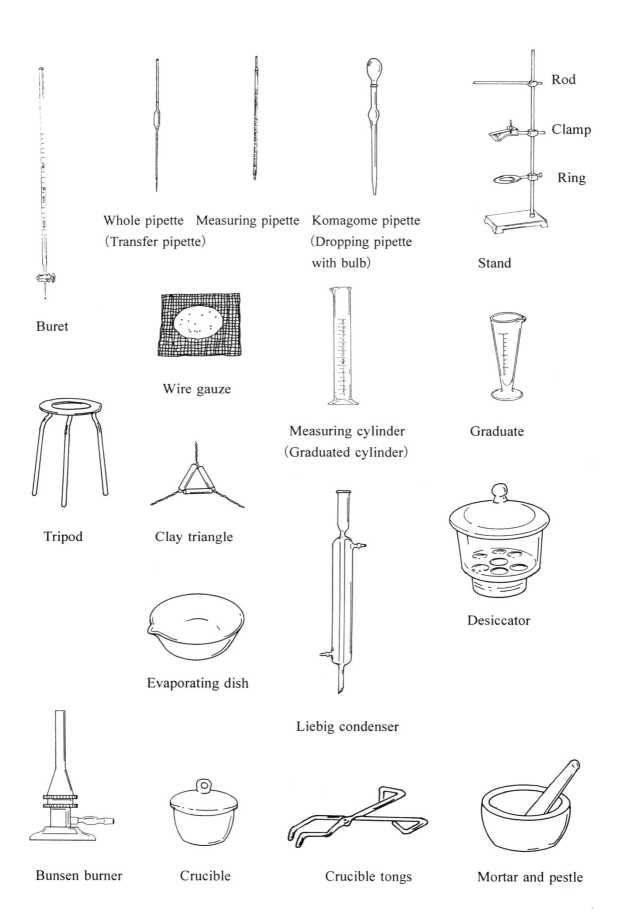

Whole pipette   Measuring pipette   Komagome pipette   Stand
(Transfer pipette)   (Dropping pipette
                                    with bulb)

Buret

Wire gauze

Measuring cylinder   Graduate
(Graduated cylinder)

Tripod   Clay triangle

Evaporating dish

Liebig condenser

Desiccator

Bunsen burner   Crucible   Crucible tongs   Mortar and pestle

Rod
Clamp
Ring

2 Practice for Safety

2.1 Doing chemical laboratory work

The principle of safety precautions is to recognize that "An accident may happen unexpectedly", even if you are strictly obeying the instruction manuals and doing your experiment carefully. Following general instruction will be helpful throughout the experiment:

1) Always wash your hands immediately after handling chemicals.
2) Always make an effort to keep the laboratory clean. Any ware and/or chemicals which are unnecessary for the performing experiment should be removed from the laboratory work bench.
3) In the case where water facilities are insufficient or unavailable, set a bucket full of water beside the work bench.
4) Even if a fire extinguisher is equipped, prepare a bucket filled with dry sand.

The illustrations shown below are for your easy understanding of the more detailed points concerned with doing chemical laboratory work.

Wear protective goggles over your eyes when handling chemicals (in particular, liquids). Keep standing when working with liquids. Hold a bottle with its label-pasted side upward so as to avoid damaging the label with chemicals. Let liquid chemicals flow down on a glass stick surface when introducing them into a beaker.

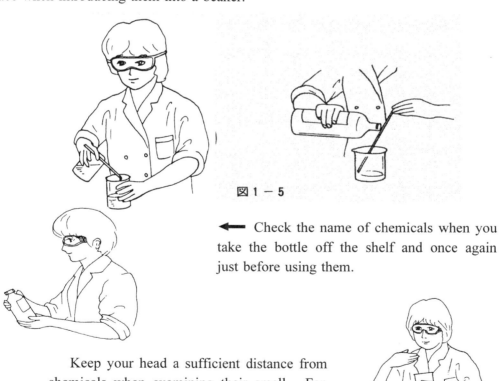

図 1 - 5

← Check the name of chemicals when you take the bottle off the shelf and once again just before using them.

Keep your head a sufficient distance from chemicals when examining their smell. Fan the smelling air with your hand. ➡

7

 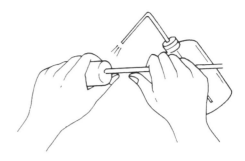

Do not insert a glass tubing into a rubber stopper with force.   Get the glass tubing and/or stopper wet so as to reduce the friction between the two.   Wrap the glass tubing in a towel and insert while slowly twisting the stopper.   Your hands should be as close together as possible.

2.2 Chemical hazards

Chemicals should be handled with the utmost caution when being used, stored or disposed. Chemicals should be handled under the direction of qualified individuals familiar with their potential hazards and trained in proper laboratory procedures.   Know the symbol marks （pictographs） for chemical hazards for safety.   Hazard warnings are listed for products in catalog and on the product label.   Watch out for them on bottles and other chemical containers. They are given below:

 EXPLOSIVES

Danger: Fire, blast or projection hazard

Examples: Mixture of C, S, KNO_3 and resin（black powder）, $C_6H_2(OH)(NO_2)_3$ （picric acid）

 FLAMMABLE GASES and AEROSOLS

Danger: Extremely or highly flammable

Examples: CH_3OH, C_2H_5OH, C_6H_{14} （hexane）

 GASES UNDER PRESSURE

Danger: Explosion or cryogenic burns

Examples: H_2, CO_2

 OXIDIZING GASES, LIQUIDS and SOLIDS

Danger: Explosion or fire

Examples: H_2O_2, $KMnO_4$

ACUTE TOXICITY

Danger: Fatal or toxic if swallowed, inhaled
or in contact with skin
Examples: Cl_2, H_2S, Hg

CORROSIVE
to METALS,
SKIN CORROSION
or IRRITATION

Danger: Skin burns,
skin/eye damage/irritation
Example: *Conc.* H_2SO_4

GERM CELL
MUTAGENICITY,
CARCINOGENICITY

Example: C_6H_6 (benzene)

AQUATIC TOXICITY

Notable pollutants: Greenhouse gases,
heavy metals, volatile organic compounds,
polycyclic aromatic hydrocarbons

Don't just throw everything into the sink. All chemicals should be treated for proper disposal procedures to prevent environmental pollution.

3 Basic Techniques in Laboratory

3.1 Use of pipette

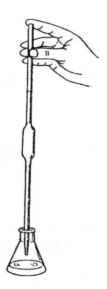

When using a pipette, do not use your mouth to take out solutions of strong acid/alkali or organic solvents. Use a pipetter or some other appropriate tools. The illustration shown above is a proper way to handle a whole pipette.

3.2 Preparation of Gases

Preparation of gases in a laboratory is characterized by the following three points:

1) Amount of gas needed is small.

2) Manipulation is easy and the purity of the gas evolved is high.

3) Preparation of gases in a laboratory is far more expensive than the preparation of those produced on the industrial scale.   However, cost is a minor factor for the gases treated in this manual.   Gases evolved are collected in three different ways:

"Over-water" method

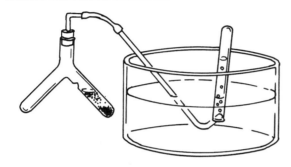

For gases whose solubility in water is low, the gases evolved are collected by substituting the water in a test tube (or a gas collecting bottle) with the gases.   Examples of application of this method are H_2, O_2, C_2H_2, and C_2H_4.

"Downward-delivery" method

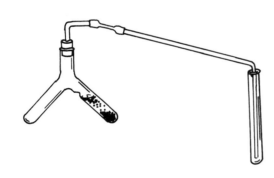

For gases whose solubility in water is high, "Over-water" method is not applicable.   Among such gases, if the gases are heavier than air, the evolved gases are collected in an unsealed bottle or test tube.   This can be seen in the illustration on the left.   By this method, the gases gradually occupy the space in the bottom of the bottle. Examples are Cl_2, HCl, SO_2, and CO_2.

"Upward-delivery" method

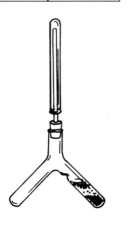

For gases which have high solubility in water and densities smaller than that of air, a gas-collecting test tube or a bottle is placed with its mouth directed downward over the glass tubing. The typical example of application of this method is the case of NH_3.

Part II Experiments

In Part II that follows, 12 experimental themes in total that start from "Conservation of elements" will be presented. Thus we ask you to think again about the history of elements at this point of start working on experiments, though you may be familiar with what the term element means.

From ancient times to today, how elements that are elemental components of all substances were discovered and recognized is an important part of human cultural history. Even in the Greek period, some materials such as gold and silver were regarded as elements. The diagram below shows how the number of elements changed over the years. By the end of the 19th century, about 80 kinds of elements were discovered. In the early half of the 20th century, the concept of elements was forced to change radically as a result of discovery of phenomena of radioactive decay of elements. Then appeared an accelerator that "can synthesize new elements", which lead to the synthesis of trans-uranium elements.

Today, the elements with the atomic number up to 118 are confirmed.

The 113th element Nihonium（Nh）was synthesized in Japan and the name reflects that the phonetic expression of Japan（日本）in Latin characters is "Nihon."

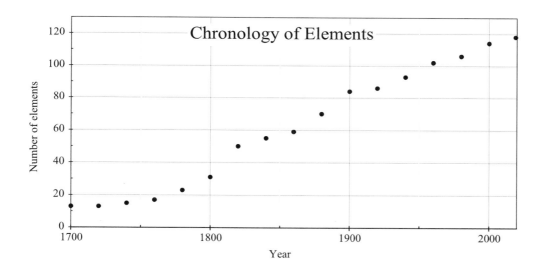

Photographs

These photographs are indicated as
" ☞ *See* Photograph " in Part II .

① Water purified by reverse osmosis (p.13)

② Hygroscopic nature of NaOH (p.32)

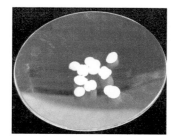

NaOH taken from the reagent bottle

⇩

NaOH after absorbing moisture in the air

Sodium hydroxide is highly hygroscopic and readily dissolves in the water it absorbs.

③ End point of titration (p.37)

The reaction mixture turns red-purple at the end point of neutralization titration. The paler the color of the solution is, the better the results of experiments will be. This color tends to fade by the absorption of carbon dioxide in the air.

④ Reaction of Cu^{2+} (p.44)

Cu^{2+}, $Cu(OH)_2$
$[Cu(NH_3)_4]^{2+}$, CuS
(From left to right)

5 Reaction of Ag⁺ (p.45)

Ag⁺, Ag₂O, [Ag(NH₃)₂]⁺, Ag₂S
(From left to right)

6 Silver halides (p.45)

AgCl, AgBr, AgI
(From left to right)

7 Silver-mirror reaction (p.53)	8 Color of phenol (p.56)

Silver-mirror reaction is used to make many kinds of mirrors.

Phenol is colorless, but it tends to show pale red color according to aging.

9 Fehling's solution and its reaction (p.60)	10 Reaction of starch with iodine (p.61)

Fehling's solution (left) and the change (right) after adding reducing sugar are shown. The product of the reaction is Cu_2O (reddish brown).

Starch-iodine reactions are different depending on the kinds of starches.

Basic operations — Conservation of elements

Purpose

All material things on earth are made up of one or more elements. Elements are substances that cannot be split by ordinary chemical methods into simpler substances. In the *Periodic table* of elements, they are listed in order of their atomic numbers. Confirm that even if the colors and the properties of a substance change by chemical reaction, elements in the substance are conserved. In addition, learn basic operations in chemistry laboratory.

Keywords

Substance, Element, Chemical reaction, Filtration, Chemical equation

Preparation

［Ware］　□ Evaporating dish　　　□ Test tube

□ Glass stick　　　　　　□ Measuring cylinder（20 cm^3）

□ Spatula　　　　　　　□ Filter paper

□ Funnel　　　　　　　□ Weighing paper

□ Funnel stand　　　　　□ Forceps

□ Heating tools（Bunsen burner, Tripod, Wire gauze）

［Reagents］　□ Cu（copper, powder）　□ Fe（iron, nail）

□ 1mol/L H$_2$SO$_4$（sulfuric acid）

Procedure

I　Basic operations

〈Purified water〉

Purified water is indispensable for experiments. Very often, however, purified (distilled) water is unavailable in developing countries. In such cases, use of commercially available drinking water, preferably prepared by *reverse osmosis*, is recommended.

☞ **See** Photograph ①

Additional Information：Osmotic pressure（π）is the pressure exerted by the solvent molecules permeating through a semi-permeable membrane to more-concentrated solution. Reverse osmosis is a method to prepare water from saline water. In this method, an excess pressure over π is applied to the solution side in the system consisting of solvent side and solution side separated by a semi-permeable membrane.

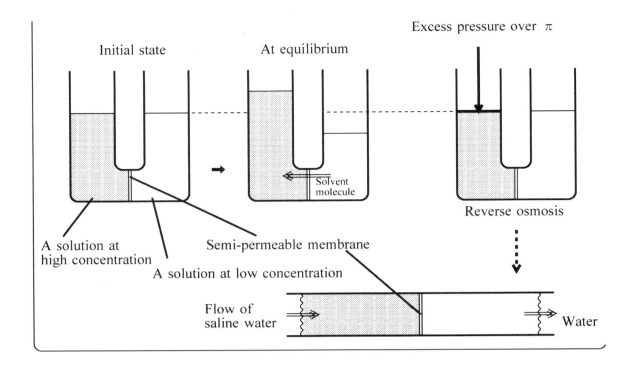

A solution at high concentration

A solution at low concentration

Semi-permeable membrane

Flow of saline water

Water

<Proper way to handle the Bunsen burner>

1) Close softly both the air adjustment screw (a) and the gas adjustment screw (b). Next, open the main valve of gas line and the valve of the burner.

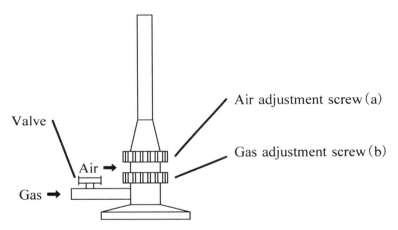

Valve

Air adjustment screw (a)

Gas adjustment screw (b)

Air

Gas

2) Do not light the burner from above. Place a lighter (or a glowing splint) close to the top of the burner stem and light from the side. Fire the gas by opening (b) little by little.

3) Turn (a) with one hand to introduce an appropriate amount of air that will make a light blue flame, while fixing (b) with the other hand. Don't look down at the burner as you open the valve.

4) To put out the fire, turn (a) and (b) together and close. Next, close the main valve and the valve of the burner.

〈Proper way to take liquid reagent from a bottle〉

Hold a reagent bottle with the label upward and take the plug out of the bottle with the other hand. Pour the reagent into a test tube. You should not put the plug on the table.

〈Proper way to measure the volume of liquid〉

The surface of a liquid in a measuring cylinder is concave. The correct volume of liquid is given by the base of the surface. Read the meniscus to 1/10 of the smallest division.

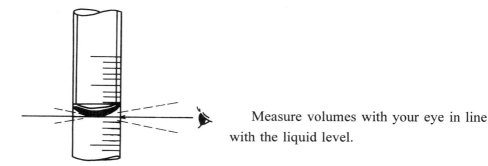

Measure volumes with your eye in line with the liquid level.

II　Heating of copper powder

1　About 1 g of Cu powder is placed in an evaporating dish. Heat it sufficiently (about 2 minutes) until a change is observed, while stirring the Cu powder with a glass stick. The fresh surface of solid copper is slightly shinny and reddish-brown.

・How has the color of Cu powder changed by heating?

・What change has taken place by the reaction?

III　Reaction of the copper product with dilute sulfuric acid

2　After the contents of the evaporating dish used are cooled, transfer them onto a weighing paper, then introduce them into a test tube. Next, add 10 cm^3 of 1mol/L H_2SO_4 to the test tube and heat the mixture.

・How does the solid change?

・How does the solution change?

15

IV   Reaction of iron nail in the solution containing copper ions

3   Separate the solid by filtration.   Add a new iron nail to the filtrate.

　　· What kind of substance is included in the filtrate?

　　· What kind of change has taken place on the surface of the iron

　nail?

〈Proper way to handle filter paper and to filtrate〉

1） Fold up a piece of filter paper twice and open to a cone.   Put it in a

　funnel.

2） Wet the filter paper with water and stick it to the funnel.

3） During the filtration, put the exit of the funnel against the wall of

　the vessel.

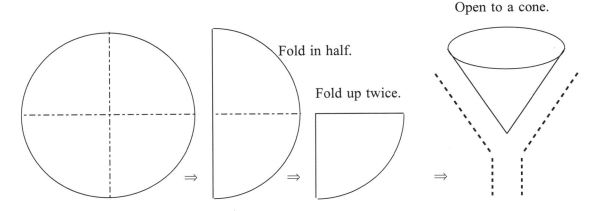

Open to a cone.

Fold in half.

Fold up twice.

Attention and Tip to overcome difficulties in the experiment :   If no
filter paper is found, use commercial coffee filter paper.   In this
case, if the filtrate is still turbid, then repeat the filtration with the
filter paper used just before.   Treat the residue on the filter paper as
waste following the instructions of your teacher.

Questions for students

1   Write the equations for the reaction you have observed in Procedures 1 and 2.

2   Write the equation for the replacement reaction occurred on the surface of the iron nail.

Chemical equation
— Confirmation of the amount of substance

Purpose

Chemical equations are used to represent chemical reactions. In a chemical reaction, the starting substances are called *reactants* and the substances formed *products*. Chemical equations show what changes take place and also the relative amounts of the compounds that take part in the reactions. Consider the case of the reaction to generate CO_2 gas. In the laboratory, CO_2 gas is generated in most cases by the reaction:

$$CaCO_3 + 2HCl \rightarrow CaCl_2 + CO_2 + H_2O$$

The above equation shows that 1 mol of the CO_2 gas arises from 1 mol of $CaCO_3$. The molar mass of $CaCO_3$ is _____ g/mol. Thus, 10 g of $CaCO_3$ produces about 2.2 L of CO_2 gas. Measure the amount of CO_2 gas evolved. Confirm the above relation between the amount of the reactant $CaCO_3$ and the product CO_2 on the basis of the equation of state of gas.

Keywords

Chemical equation, Reactant, Product,

Measurement of gas volume, Vapor pressure,

Equation of state of gas (For detail, see p.21.)

Preparation

[Ware] □ Polyethylene bag (about 1 L, 20 cm × 30 cm)

□ Glass tubing (about 10 cm, fitted with a rubber tubing (5 - 7 cm))

□ Measuring cylinder (1 L)

□ Trough

□ Thermometer

□ Barometer

[Reagents] □ *conc*. HCl (concentrated hydrochloric acid)

□ $CaCO_3$ (calcium carbonate)

> Tip to overcome difficulties in the experiment : If $CaCO_3$ as a chemical reagent is unavailable, well-washed and dried seashell or eggshell is usable. Use of marble stone is unsuccessful for this experiment.

Procedure

    I   Generation of CO_2 gas

1 Take about 5 g of $CaCO_3$ and measure the weight to first decimal place. If you use seashell or eggshell, you should crack them into pieces.

    · Amount of $CaCO_3$ used: _____ g = _____ mol ···①

    · Theoretical amount of HCl you need:

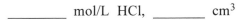

                           _____ mol/L HCl, _____ cm^3

2 Introduce $CaCO_3$ and an appropriate amount of *conc.* HCl into the bag as shown below. The amount of HCl is slightly in excess of the theoretical one.

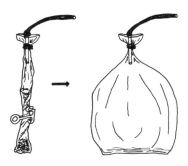

3 Dissolve $CaCO_3$ in the bag with *conc.* HCl. The bag expands with the dissolution of $CaCO_3$. In the case where a significant amount of shells remains, weigh the shells left after the completion of the reaction and correct the value ①.

    · Amount of HCl poured:

        _____ cm^3 of _____ mol/L HCl = _____ mol of HCl

4 Squeeze the evolved CO_2 gas out of the bag. Collect the gas by the "Over-water" method (see p.10) using a measuring cylinder. In order to prevent dissolving any CO_2 into the water, you had better add 1 cm^3 of 1mol/L HCl per 10 L of water in a trough.

> Caution :   Be careful when squeezing the CO_2 gas out of the bag and into the measuring cylinder, for the bag contains HCl! Put on protective gloves.

II   Measuring the volume of the CO₂ gas

5   Measure the volume of CO₂ gas collected in the cylinder.   In the measurement, the water surface in the cylinder should be at the same level as the water surface in the trough.

   · Volume of CO₂ gas evolved: _____ cm^3

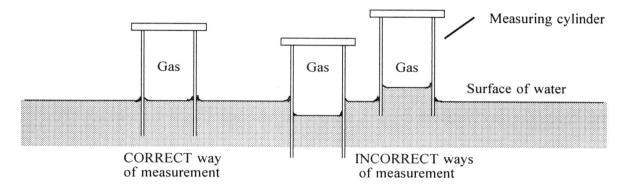

CORRECT way   INCORRECT ways
of measurement   of measurement

6   Record the water temperature in the trough, room temperature, and the atmospheric pressure.

   · Water temperature: _____ ℃   Room temperature: _____ ℃
   · Atmospheric pressure: _____ hPa = _____ mmHg
   (If the barometer is unavailable, assume the atmospheric pressure depending on the weather to be at 1020 hPa on a fine day, and 1000 hPa on a rainy day.)

Additional Information :   Note that the volume of the bag measured in this procedure refers to the volume occupied not only by CO₂ molecules but also H₂O molecules, since water vapor also comes into the bag with CO₂. As will be explained on p.21, the external pressure (*i.e.* atmospheric pressure p_{atm}) is balanced with the sum of the partial pressure for CO₂, $p_{(CO_2)}$, and that of H₂O, $p_{(H_2O)}$, namely:

$$p_{atm} = p_{(CO_2)} + p_{(H_2O)}$$

   The vapor pressure of water is given in **Table 1** below.

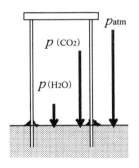

Table 1: Vapor pressure〔mmHg〕 of water

℃	0	1	2	3	4	5	6	7	8	9
0	4.6	4.9	5.3	5.6	6.1	6.5	7.0	7.4	8.1	8.6
10	9.2	9.8	10.5	11.1	12.0	12.7	13.6	14.4	15.5	16.4
20	17.5	18.5	19.8	20.9	22.3	23.6	25.2	26.6	28.4	29.8
30	31.6	33.5	35.4	37.4	39.6	41.9	44.3	46.7	49.4	52.1

   1013 hPa〔hPa: hectopascal〕 = 1 atm = 760 mmHg

Questions for students

1 Explain the origin of naturally occurring $CaCO_3$ on the earth crust including oceanographic environment.

2 The CO_2 content in the earth atmosphere attracts much attention. What methods are used to know the CO_2 content in the air?

3 If H_2SO_4 is used in place of HCl in this experiment, the gas CO_2 will not be generated easily. Explain why the reaction to generate CO_2 will not proceed smoothly with H_2SO_4.

4 In Procedure 5, why should the water surfaces inside and outside of the measuring cylinder be at the same level in order to measure the volume of CO_2 gas? In reality, however, the error coming from this factor does not affect the result. Why?

5 Calculate the amount (in mol) ② of CO_2 evolved in this reaction. Use the equation of state of gas (see p.21) $p_{(CO_2)} V = nRT$, where $p_{(CO_2)}$ is the partial pressure of CO_2, V the volume of the bag, and T the temperature in Kelvin. Adopt the average temperature of ambient and water temperatures. Then, compare the ratio ②/①.

6 When $1 cm^3$ of 1mol/L HCl is added into 10 L of water, what is the approximate value of the pH of the water (i.e. an extremely dilute solution of HCl)? The pH is defined as the relation $pH = -\log [H^+]$, where $[H^+]$ stands for the concentration [mol/L] of H^+ ion.

7 Discuss the factors that may have affected the accuracy of your result, and answer how you will change the procedure to get more accurate results.

Molecular weight of gases

Purpose

Each gas has its own mass. Thus, the mass of gas depends on chemical species. Calculate molecular weight from the mass of a known volume of gas. A gas collected by "Over-water" method (explained on p.10), contains water vapor. Learn that the partial pressure exerted by the water vapor must be accounted for in those cases.

— Equation of state of gas : The equation $pV = nRT$ is called *equation of state of gas*, where $p[\text{Pa}]$ is the pressure, $V[\text{m}^3]$ the volume, $n[\text{mol}]$ the amount of substance, $T[\text{K}]$ the temperature in Kelvin, and R the gas constant $8.31[\text{J}/(\text{mol}\cdot\text{K})]$.

— Law of partial pressure : When a gas is a mixture, the pressure each component gas exerts is called its *partial pressure*. Therefore, the *total pressure* of a gas collected over water is the sum of the partial pressure exerted by the gas and the partial pressure exerted by water vapor.

— Avogadro's law : The number of gas molecules in a container determines the pressure at a given temperature. Avogadro's law states that under the same conditions (temperature and pressure), equal volumes of gases contain an equal number of molecules. Molar volume of ideal gas is 22.4 L at 273.15 K, 101325 Pa.

Keywords

Molecular weight, Gas, Equation of state of gas, Partial pressure,
Total pressure, Atmospheric pressure, Avogadro's law

Preparation

[Ware] □ Rubber band   □ Trough

  □ Pinch cock   □ Bucket

  □ Measuring cylinder (1 L)   □ Barometer

  □ Balance (measurable up to 0.01 g scale)

  □ Thermometer (0 - 100 ℃, 1/2 degree scale)

  □ Polyethylene bag (about 1 L, 20 cm × 30 cm)

  □ Rubber plug (Open a hole and make a ditch to set rubber band.)

  □ Rubber tubing (50 - 60 cm in length and connected to a glass tubing)

  □ Glass tubing (5 - 7 cm in length, to one end of which a rubber tubing
    (5 - 7 cm in length) is attached)

[Reagents] □ O_2 (oxygen, cylinder)

  □ CO_2 (carbon dioxide, cylinder)

Forecast

Does gas have mass?   Your reply is _____.

Do gases of the same volume have the same mass irrespective of the kinds?   Your reply is _____.

Procedure

I   Measuring the mass of CO_2

1   Make a device from a polyethylene bag to put gas in.   Be sure that the larger diameter side of the plug should be directed to the inside of the bag.   Push out the air from the device and close the rubber tubing with a pinch cock and weigh the mass of the whole device to 0.01 g.

Mass of the whole device _____ g ···①

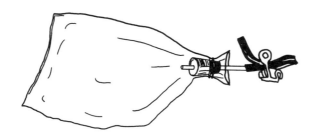

2   Connect the device to a CO_2 cylinder. Expand the bag sufficiently with CO_2 gas.   Next, place the rubber plug lightly so that the plug is on the upper side of the expanded bag.   Loosen the pinch cock and let the pressure in the bag be equal to the atmospheric pressure. Close the pinch cock and weigh the mass of the whole device to 0.01 g.

Mass of the whole device containing CO_2 _____ g ···②

II   Measuring the mass of O_2

3   Push out CO_2 from the bag.   Fill the bag with O_2.   Weigh the mass of the whole device following the procedure 2.

Mass of the whole device containing O_2 _____ g ···③

III   Measuring the volume of the bag

4   Fill a measuring cylinder with water and stand it upside down in a trough.   Collect the O_2 gas used in Procedure 3 in the measuring cylinder by pushing it out from the bag.   In this gas collection, use another rubber tubing if necessary.   Pay attention to the back-suction and draw out the rubber tubing.   Next, take the measuring cylinder out from the trough without spilling water and stand it up on the table.   Record the volume of the bag which is equal to that of the empty space in the cylinder.

Volume of the bag _____ cm³ ···④

Additional Information :   Note that the volume of the bag measured in this procedure refers to the volume occupied not only by O_2 molecules but also H_2O molecules, since water vapor also comes into the bag with O_2.   The external pressure (*i.e.* atmospheric pressure p_{atm}) is balanced with the sum of the partial pressure for O_2, $p_{(O_2)}$, and that of H_2O, $p_{(H_2O)}$, namely: $p_{atm} = p_{(O_2)} + p_{(H_2O)}$
The vapor pressure of water is given in **Table 1** on p.19.

5   Record the water temperature in the trough, the room temperature, and the atmospheric pressure.
   · Water temperature: _____ ℃
   · Room temperature: _____ ℃
   · Atmospheric pressure: _____ hPa ＝_____ mmHg
   (If no barometer is usable, see p.19.)

IV   Supplementary experimental work

6   Stand two candles different in height in a small bucket.   Light both the candles and gently "pour" the CO_2 gas contained in the measuring cylinder into the bucket.
   · Explain the phenomena you have observed. (Think of the volume of CO_2 gas poured and that of the bucket.)

Questions for students

1 Calculate the following values from the measured values.

    ⑤ : Apparent mass of CO_2 in the bag   ②−①=_____ g

    ⑥ : Apparent mass of O_2 in the bag   ③−①=_____ g

    ⑦ : Mass of air filling the same volume of the bag   _____ g

       (from ④ and **Table 2**)

    Actual mass of CO_2 in the bag   ⑤＋⑦=_____ g

    Actual mass of O_2 in the bag   ⑥＋⑦=_____ g

    Why should the mass of the air ⑦ be added to calculate the actual mass of gas?

2 For the same volume, how many times heavier is the mass of CO_2 than the mass of O_2? This is the specific gravity of CO_2 relative to O_2. Suppose that Avogadro's law holds for the value of this specific gravity, what does the value mean?

3 Calculate the molecular weight of CO_2 from its specific gravity to O_2. Assume the molecular weight of O_2 is 32.

4 Calculate the molecular weight of O_2 using the equation of state of gas. Use the equation of state of gas: $p_{(O_2)} V = nRT = wRT/M$, where $p_{(O_2)}$ is the partial pressure of O_2, V the volume of the bag, M the molecular weight, w the actual mass of O_2, and T the temperature in Kelvin. The vapor pressure of water is in **Table 1** on p.19.

Table 2: Mass of the air [g/L]

mmHg \ °C	10	15	20	25	30	35
700	1. 14	1. 12	1. 10	1. 08	1. 07	1. 05
710	1. 16	1. 14	1. 12	1. 10	1. 08	1. 07
720	1. 18	1. 15	1. 13	1. 12	1. 10	1. 08
730	1. 19	1. 17	1. 15	1. 13	1. 11	1. 10
740	1. 21	1. 19	1. 17	1. 15	1. 13	1. 11
750	1. 22	1. 20	1. 18	1. 16	1. 14	1. 13
760	1. 24	1. 22	1. 20	1. 18	1. 16	1. 14
770	1. 26	1. 23	1. 21	1. 19	1. 17	1. 16
780	1. 27	1. 25	1. 23	1. 21	1. 19	1. 17
790	1. 29	1. 25	1. 23	1. 21	1. 20	1. 19

1013 hPa ＝ 1 atm ＝ 760 mmHg

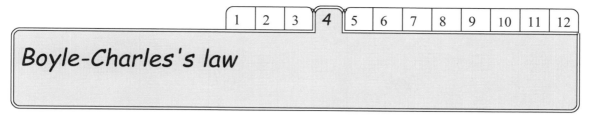
Boyle-Charles's law

Purpose

The British chemist, Robert Boyle (1627 - 1691) arrived at the classical principle on the pressure and the volume of a gas through his experiments in 1662. At constant temperature, pressure varies inversely with volume. The product of pressure and volume is then a constant. On the other hand, changes in temperature have a greater effect on the volume of gas than for the case of liquid or solid. The French physicist, Jacques Charles (1746 - 1823) noticed a simple relationship between the volume of gas and temperature in 1787; for each Celsius degree increase in temperature, the volume of gas increases by 1/273 of its volume at 0 ℃. Later, the absolute temperature was defined and the two laws on the properties of gas were unified as *Boyle-Charles's law* expressed as pV/T is constant. Confirm Boyle-Charles's law.

- Boyle's law : If the temperature of a gas remains constant, pressure exerted by the gas varies inversely with the volume. In a mathematical formula $pV = k$, where p is the pressure, V the volume, and k a constant.

- Charles's law : The volume of a fixed quantity of gas, held at constant pressure, varies directly with the absolute temperature. In a mathematical formula $V/T = k'$, where V is the volume, T the absolute temperature, and k' a constant.

Keywords

Gas, Pressure, Volume, Absolute temperature,
Boyle's law, Charles's law, Boyle-Charles's law

Preparation

[Ware]　□ Injector (or Enema, 20 cm³)　　□ Stand
　　□ Ruler　　　　　　　　　　　　　□ Beaker (1 L)
　　□ Rubber plug (or Eraser)　　　　　□ Measuring cylinder (25 cm³)
　　□ Balance (measurable up to ca. 8 kg)　□ Trough
　　□ Rubber tubing　　　　　　　　　□ Test tube clamp
　　□ Heating tools　　　　　　　　　□ Thermometer (0 - 100 ℃, 2 pieces)
　　□ Test tube (15 mm or more in diameter)　□ Slide gauge
　　□ Rubber plug fitted with L-shape glass tubing

Procedure

I   Inspection of Boyle's law

1   Measure with a ruler the length ① between the graduations 0 and 20.0 cm³ of an injector.   _____ cm   ···①

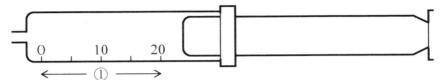

Next, calculate the cross-section area ② of the injector.

$$20 \text{ cm}^3 \div ① = \underline{\hspace{2cm}} \text{ cm}^2 = \underline{\hspace{2cm}} \text{ m}^2 \quad ···②$$

Alternatively, the cross-section area can directly be calculated from the diameter of the injector piston.

2   Measure the atmospheric pressure.

$$\underline{\hspace{2cm}} \text{ hPa} = \underline{\hspace{2cm}} \text{ Pa} \quad ···③$$

(If no barometer is usable, see p.19.)

3   Make a hole on a rubber plug (or an eraser) so that the tip of the injector just fits tightly.

4   Fix the piston at the position of 20.0 cm³ by inserting the tip to the hole of the rubber plug and load the injector on the balance plate as shown below.   The reading of the balance at this moment is taken as 0.0 kg and pressure of the air in the injector is ③.

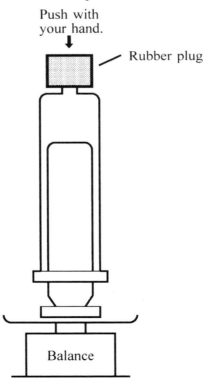

Push with your hand.

Rubber plug

Balance

5   Push the rubber plug with your hand to compress the air in the injector to 15.0 cm³, and read the graduation of the balance. The difference A[kg] in readings before and after compression is converted into the corresponding pressure B[Pa] as follows:

$$B = A \times 9.8 \div ② = \underline{\hspace{2cm}} \text{ Pa} \quad \cdots ④$$

where 9.8 ms⁻² refers to the constant of gravitational acceleration. The pressure of the air inside the injector p[Pa] is the sum of the pressures ③ and ④.

$$p = ③ + ④ = \underline{\hspace{2cm}} \text{ Pa} \quad \cdots ⑤$$

6   Repeat Procedure 5 to read each graduation of the balance when the piston is pushed to 12.5 cm³, 10.0 cm³, and 7.5 cm³. Repeat this compression process at least three times. Take the averages to calculate each pressure of the air in the injector.

Development

For the case of volume expansion, do as follows. Hang the injector with a rubber plug upward, and suspend a weight/weights to pull down the piston as shown in the figure below. Be sure to tightly bind the rubber plug and the injector.

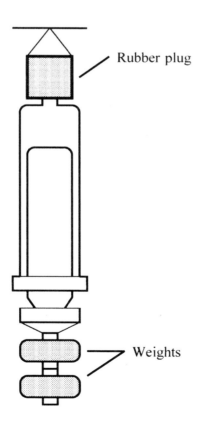

Rubber plug

Weights

II Inspection of Charles's law

7 Prepare two test tubes, which are referred to as test tube (a) and (b), respectively. Dry the inside of test tube (a) well, and attach a rubber plug with L-shape glass tubing connected to a rubber tubing (30 - 40 cm). Confirm that the air does not leak from the rubber tubing.

8 Place water in a trough. Fill a measuring cylinder with water and stand it upside down as shown in the figure below and fix it with a stand.

9 Record the room temperature t_1 with the precision 0.1 ℃. Pay attention in order that your body temperature and breath do not affect the air temperature in the tube.

10 Place about 500 cm³ of water in a 1 L beaker and heat the water to boil. Next, stop heating and add cold water to the boiling water in the beaker to set temperature of water in the range of 40 - 80 ℃. Immerse the test tube (b) fitted with a thermometer into the hot water. Record the temperature of the water in the beaker and the temperature of the air t_2 in test tube (b). Test tube (b) acts to monitor the temperature of the sample air during measurement.

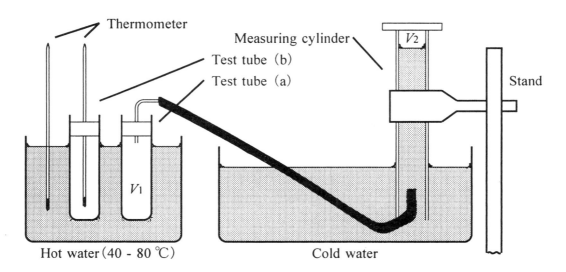

Hot water (40 - 80 ℃)   Cold water

11 Into the measuring cylinder insert the tip of the rubber tubing connected to test tube (a). Hold test tube (a) with a test tube clamp and put test tube (a) into the hot water. The rubber tubing, however, should be outside the hot water. The air bubbles come out from the rubber tubing and enter into the measuring cylinder. Keep the two test tubes in the hot water for several minutes. Air temperatures in test tubes (a) and (b) are assumed to be the same.

12 Record the temperature of hot water t_3 again. Water temperature will drop during the measurement. Then, take an average t_4 of the temperatures measured before and after Procedure 11 as the water temperature in the measurement. Measure the volume of the air V_2 that has entered into the measuring cylinder. (For reading the graduation of measuring cylinder, follow the notice on p.15 〈Proper way to measure the amount of liquid〉.)

13 Repeat Procedures 10 - 12 at several different temperatures.

14 Take out test tube （a） and introduce water until it just touches the lower surface of the rubber plug. Next, transfer all the water to the measuring cylinder to measure its volume. This is the volume of sample air V_1 in the test tube （a）.

Questions for students

1 Record the results of Procedures 4 - 6 and calculate the values of pV in the next table. Next, draw the graphs expressing the relationships between p and V, as well as pV and V.

Volume of air $V[\text{m}^3]$	20.0	15.0	12.5	10.0	7.5
Reading of balance［kg］					
Difference［kg］	0				
$p[\text{Pa}]=$ ⑤	③				
$pV[\text{J}]$					

2 From the graphs drawn in 1, summarize the relationship between the pressure and the volume of a gas. Does Boyle's law hold in your experimental results?

3 In the calculation of air pressure in the injector in Procedure 5, why is ③ added to the readings of balance?

4 To confirm the validity of Boyle's law with your data alone, the plot p against $1/V$ would be more appropriate than the plot of p against V. Plot your data as $1/V$ - p relation and explain why the former plot is better than the latter one.

5   Record the results of Procedures 11 - 14 in the next table.   Next, draw a graph expressing the relationship between t_4 and $(V_1 + V_2)$

Room temperature t_1 [℃]					
Water temperature at the onset of experiment t_2 [℃]					
Water temperature at the completion of experiment t_3 [℃]					
Average water temperature $t_4 = (t_2 + t_3)/2$ [℃]					
Difference in temperature $\Delta t = t_4 - t_1$ [K]					
Original volume of sample air V_1 [cm^3]					
Volume of the air that has expanded V_2 [cm^3]					
Volume expansion ratio $(V_2/V_1)/\Delta t$					

6   From the graph drawn in 4, summarize the relationship between the temperature and the volume of a gas.

7   Why is the immersion of the rubber tubing in hot water improper in Procedure 11?

8   The volume expansion ratio of air is $0.0037 (= 1/273)$ K^{-1} at 0 ℃, 1013 hPa.   Calculate the error in your measurement from the average value of your results.

9   Charles's law is expressed alternatively as follows:
$$V = V_0(1 + t/273) = V_0(273 + t)/273$$
where V_0 and V are the volumes of gas at temperatures 0 ℃ and t ℃.   Derive this equation from the relation $V/T = k'$ given on p.25 and prove that the volume expansion ratio is 0.0037 $(= 1/273)$ K^{-1}.

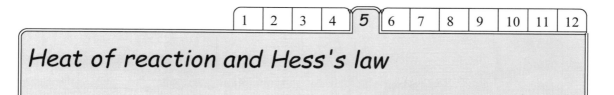

Heat of reaction and Hess's law

Purpose

Most chemical reactions are accompanied by heat. The heat is the reflection of bond breaking and bond formation in the reaction. If heat is evolved, the reaction is called *exothermic*, while if heat is absorbed, the reaction is said to be *endothermic*. The reaction formula in which the amount of heat change is explicitly expressed is called *thermochemical equation*. Measure the heat of solution of sodium hydroxide, heat of neutralization of hydrochloric acid with sodium hydroxide (solid), and heat of neutralization of hydrochloric acid with sodium hydroxide (solution). Confirm the validity of Hess's law.

— Hess's law : The enthalpy change of a reaction under consideration is the sum of the enthalpy changes of a series of reactions that add up to that reaction.

Additional Information : Precise definition of enthalpy requires understanding of thermodynamics. Here, readers can take enthalpy change as heat energy accompanying the reaction.

Keywords

Heat of reaction, Exothermic reaction, Endothermic reaction,

Heat of solution, Heat of neutralization,

Hess's law, Thermochemical equation

Preparation

[Ware]   □ Polystyrene foam cup (200 cm^3, 3 pieces)

□ Stirring stick

□ Measuring cylinder (100 cm^3)

□ Thermometer (0.1 or 0.2-degree scale)

□ Balance

□ Spatula

[Reagents]   □ 1mol/L HCl (hydrochloric acid)   □ 2mol/L HCl

□ 2mol/L NaOH (sodium hydroxide)   □ NaOH (solid)

Caution : Prior to experiments, leave the water and the sample solution for a while so that their temperatures are in equilibrium with the room temperature.

Procedure

I   Heat of solution of NaOH(s)

1   Place 100 cm³ of water and a thermometer in a polystyrene foam cup. Measure the temperature for about 5 minutes at intervals of 30 seconds. Record the temperature to 1/10 of the smallest division.

> Caution : Weigh NaOH(s) quickly, because NaOH is highly hygroscopic.   ☞ **See** Photograph ②

2   Take about 4.0 g of NaOH(s) with precision up to 0.1 g unit and put it in the cup smoothly. While stirring well to dissolve the NaOH completely (several minutes will be needed.), measure the temperature for about 10 minutes at intervals of 30 seconds. Read the highest temperature from the plot of the variation of the temperature with time. In the plot, the data obtained in Procedure 1 should also be included.

> Additional Information : In an ideal condition, a rise in temperature follows the line A. The behavior observed in experiment is, however, like the one given by the line B. The highest temperature (T_1) is not T_1'. T_1 is estimated after extrapolating the line B to time t_0, as shown in figure. (t_0: Time of reaction initiation, T_0: Initial temperature of reactants)

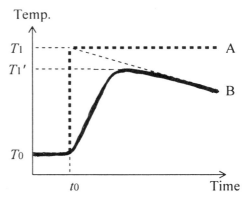

3   Measure the following values.

Mass of the cup : _____ g   Total mass : _____ g

Initial temperature : _____ ℃   Highest temperature : _____ ℃

II   Heat of neutralization of HClaq with NaOHaq (aq: aqueous solution)

4   Place 50 cm³ of 2mol/L HCl and a thermometer in another cup. Measure the temperature for about 5 minutes at intervals of 30 seconds. Record the temperature to 1/10 of the smallest division.

5   Measure the temperature of 2mol/L NaOH. Add this solution to the HCl solution treated in Procedure 4. Note that the rise in temperature is far more rapid than in Procedure 2. While stirring well, measure the temperature for about 5 minutes at intervals of 30 seconds. Read the highest temperature from the plot of the variation of the temperature with time. In the plot, the data obtained in

32

Procedure 4 should also be included.   Confirm that the resulting solution is neutral or basic.

6   Measure the following values.

Mass of the cup : _____ g   Total mass : _____ g

Initial temperature : _____ ℃   Highest temperature : _____ ℃

Ⅲ   Heat of neutralization of HClaq with NaOH(s)

7   Place 100 cm³ of 1mol/L HCl and a thermometer in another cup. Measure the temperature for about 5 minutes at intervals of 30 seconds.   Record the temperature to 1/10 of the smallest division.

8   Add a little more than about 4.0 g of NaOH(s).   While stirring well, measure the temperature for about 5 minutes at intervals of 30 seconds.   Read the highest temperature from the plot of the variation of the temperature with time.   In the plot, the data obtained in Procedure 7 should also be included.   Confirm that the resulting solution is neutral or basic.

9   Measure the following values.

Mass of the cup : _____ g   Total mass : _____ g

Initial temperature : _____ ℃   Highest temperature : _____ ℃

Ⅳ   Calculation of the amounts of heat evolved

10   Calculate each value of ① - ⑤ in Ⅰ, Ⅱ, and Ⅲ.   Value ③ is calculated as " ② × 4.2 × Δt [J]".   Here, the specific heat capacity of the solution is taken as 4.2J/(g·K).

① :   Rise in temperature Δt[K]   (difference between the highest temperature reached and the initial temperature)

② :   Mass of the reaction mixture [g]

③ :   Amount of heat produced by reaction [J]

④ :   Amount of NaOH [mol]

⑤ :   Amount of heat per mol of NaOH [kJ/mol]

	①	②	③	④	⑤
Ⅰ					$\equiv Q_1$
Ⅱ					$\equiv Q_2$
Ⅲ					$\equiv Q_3$

In the above calculation, the heat-insulation of the polystyrene foam cup is assumed to be perfect, namely, the cup is an "adiabatic" system.

33

Questions for students

1 Write the thermochemical equations for each reaction of I , II, and III.

   I

   II

   III

2 Examine Hess's law on the basis of the data you have obtained.

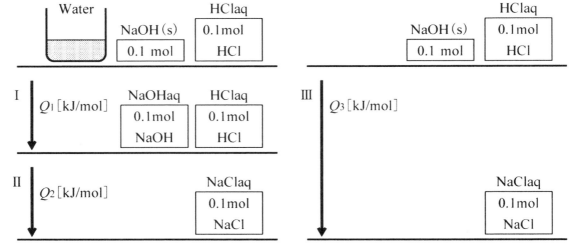

Q_1 : Heat of solution of NaOH (s) _____ kJ/mol

Q_2 : Heat of neutralization _____ kJ/mol Q_3 : Sum of heat of solution and

heat of neutralization _____ kJ/mol

Additional Information : The following values are given in literature for reactions at extremely low concentration.

   Q_1 = +44.5 kJ/mol, Q_2 = +56.5 kJ/mol, Q_3 = +101 kJ/mol

Experimental Q_2 values are often larger than the above value.

3 In Procedures 5 and 8, it is indicated to confirm that the solution is neutral or basic. Why is this checking necessary? If this is not the case, what error is contained in the values you have determined?

4 If the temperature of the HCl solution and that of the NaOH solution are different in Procedure II, how do you change the way of calculation for the heat of neutralization described in Procedure IV?

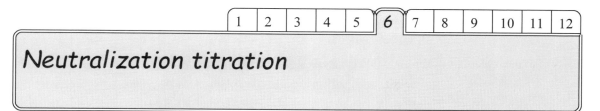

Neutralization titration

Purpose

Neutralization is the reaction between H^+ ion and OH^- ion to produce water, namely $H^+ + OH^- \rightarrow H_2O$. Neutrarization titration is one of the basic methods of volumetric analysis. Learn the concept of titration and the fundamental operation. Determine the concentrations of sodium hydroxide with neutralization titration using standard solution of oxalic acid. Then, titrate vinegar taken as acetic acid with sodium hydroxide.

Keywords

Titration, Neutralization, Standard solution, Volumetric analysis

Preparation

[Ware]　□ Buret (50 cm³)　　　　　□ Buret support

　　□ Whole pipette (10 cm³)　　　□ Conical beaker (100 cm³)

　　□ Beaker (100 cm³)　　　　　□ Funnel

　　□ Volumetric flask (100 cm³)

[Reagents]　□ 0.2mol/L NaOH (sodium hydroxide)

　　□ $H_2C_2O_4 \cdot 2H_2O$ (oxalic acid dihydrate)

　　□ Vinegar (commercial product)

　　□ Phenolphthalein solution (For preparation, see **Appendix** on p.63.)

Procedure

　I　Preparation of solutions necessary

1　Prepare 0.100mol/L $H_2C_2O_4$ standard solution as follows:

（1）Calculate the molecular mass of $H_2C_2O_4 \cdot 2H_2O$. ＿＿＿＿ g/mol …①

> Additional Information : Anhydrous oxalic acid $H_2C_2O_4$ is hygroscopic. Hydrated water molecules in $H_2C_2O_4 \cdot 2H_2O$ become a part of solvent. One molecule of $H_2C_2O_4$ contains two COOH groups, and can ionize to produce two H^+ ions.

（2）Suppose the volume of 0.100mol/L $H_2C_2O_4$ standard solution necessary to prepare is x cm³, the mass of $H_2C_2O_4 \cdot 2H_2O$ that should be taken from the reagent bottle is calculated as follows:

$$① \times 0.100 \times x/1000 = \text{＿＿＿＿} g \cdots ②$$

35

(3) Take the calculated amount ② of $H_2C_2O_4 \cdot 2H_2O$ as precisely as possible. Place it in a beaker and add water to dissolve. Next, introduce the solution quantitatively into a measuring flask and dilute it to x cm³.

2 Prepare about 0.2mol/L NaOH solution by dissolving a required amount of NaOH in water. In this case, precise weighing of NaOH and use of a volumetric flask is unnecessary.

  ・Why is the precise weighing of NaOH unnecessary? Main reasons are the following two:

————————————   ————————————

> WARNING : Be careful so as TO PROTECT YOUR EYES FROM CONTACT WITH NaOH SOLUTION.

  II Determination of the approximate concentration of NaOH solution with 0.100mol/L $H_2C_2O_4$

3 (1) Close the cock of a buret and pour NaOH solution into the buret through a funnel.

  (2) Tilt the buret to an extent that no overflowing of the NaOH solution will occur and wash the inside wall of the buret by rotating.

  (3) Open the cock of the buret and discard the NaOH solution.

4 Close the cock of the buret and clamp the buret to a buret support. Pour the NaOH solution and remove the funnel. Next, put a 100 cm³ beaker under the buret. Open the cock and pour out 2 - 3 cm³ of the NaOH solution to expel the air in the tip of the buret.

5 Take several cm³ of 0.100mol/L $H_2C_2O_4$ standard solution with a 10 cm³ whole pipette and wash the inside wall of the pipette following the same procedure as above.

6 (1) Take out the $H_2C_2O_4$ standard solution to the point above the marked line of the whole pipette (see **3.1 Use of pipette** on p.9) and with your forefinger, adjust the solution level to the marked line of the whole pipette.

  (2) Transfer the standard solution in the whole pipette to a conical beaker.

  (3) Add 1 - 2 drops of phenolphthalein solution as an indicator for titration. Wash the inside wall of the conical beaker with water.

7 Do preliminary titration as follows:

  (1) Record the reading of meniscus of the NaOH solution inside the buret to 1/10 of graduation.

36

(2) Add the NaOH solution step by step to the $H_2C_2O_4$ standard solution. In each step, add 1 cm^3 of the NaOH solution. Swirl the conical beaker until the whole mixture turns red and the color does not disappear. Evaluate the approximate quantity of the NaOH solution required. Wait one minute or so to examine whether the color remains unchanged. Then, stop putting in drops of the NaOH solution.

Ⅲ   Neutralization titration

8 (1) Considering the result of 7 (2), add the NaOH solution to the $H_2C_2O_4$ standard solution. The quantity of the NaOH solution should be smaller than the estimated end point of titration.

(2) Put in drops of the NaOH solution little by little while swirling the conical beaker gently. If the color of reaction mixture turns slightly red and the color remains unchanged even after swirling the conical beaker lightly, the mixture is at the end point of neutralization titration.   ☞ ■**See**■ Photograph ③

(3) Take a note of the quantity of the NaOH solution from the meniscus reading of the buret.

(4) Repeat the titration three times and calculate the average value.
      Quantity of the NaOH solution required
      1st _____ cm^3
      2nd _____ cm^3
      3rd _____ cm^3 Average value _____ cm^3

(5) Calculate the concentration [mol/L] of NaOH solution prepared.

Ⅳ   Quantitative analysis of the amount of acetic acid in vinegar

9   Introduce the NaOH solution into the buret whose concentration is determined in Ⅲ.

10   Take 20 cm^3 of vinegar with a whole pipette and transfer it to a volumetric flask to dilute to 100 cm^3. If a 20 cm^3 whole pipette is not available, use a 10 cm^3 whole pipette twice. Transfer 10 cm^3 of the diluted vinegar to a conical beaker and carry out titration with the NaOH solution. Repeat three times and calculate the average value.
      Quantity of the NaOH solution required
      1st _____ cm^3
      2nd _____ cm^3
      3rd _____ cm^3 Average value _____ cm^3

11   Calculate the concentration [mol/L] of acetic acid in vinegar.

37

Questions for students

1 List the properties required for reagents used for preparation of *standard solutions*. Then, confirm that $H_2C_2O_4 \cdot 2H_2O$ meets the requirements, while NaOH does not.

2 Write the equation for the neutralization of acetic acid with sodium hydroxide.

3 How many grams of acetic acid are included in 100 cm³ of the vinegar? Using the density of the commercial vinegar (1.05 g/cm³), calculate the weight percent of acetic acid in commercial vinegar.

4 Dilution of the vinegar is effective to attain high accuracy of the experiment. Why?

5 If methyl orange (MO) is used instead of phenolphthalein (PP) as an indicator in this neutralization titration, how does the result change? Find out the answer from Neutralization Titration Curve shown below.

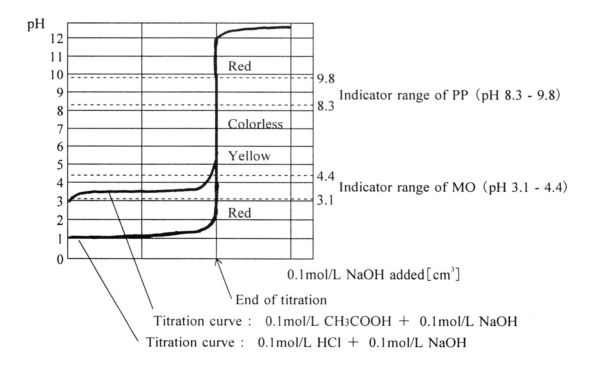

Neutralization Titration Curve

38

Preparations of some indispensable reagents from daily materials

Purpose

Chemicals are prepared from various resources and they are supplied for the chemical industry. In our environment, chemical reactions, including those between acids and bases, are very common. At this stage, think of the terms acid, alkali, and salt. Acid comes from the latin word *acidus*, meaning sour, while alkali is derived from an Arabic word meaning a substance derived from the ash of plants. By the 18th century, chemists recognized salts as the product of the reaction between an acid and an alkali. Thus, the term base was used for any substance capable of reacting with acid and serving as a *base* for the salt. There are many bases that do not have OH⁻ ions as part of their original formulation. Nowadays, alkali means bases or hydroxides such as NaOH, Ca(OH)$_2$(slaked lime), and Mg(OH)$_2$(milk of magnesia). These are soluble in water and are able to increase the concentration of OH⁻ ions in solution. Prepare several key chemicals, such as acid HCl and bases Ca(OH)$_2$, NaHCO$_3$, Na$_2$CO$_3$, NaOH, and K$_2$CO$_3$ from daily materials and understand the relations between the compounds treated.

Keywords

Acid, Alkali, Base, Ammonia soda process(Solvay soda process)

Preparation

[Ware]　☐ Flat bottom flask

☐ Erlenmeyer flask

☐ Beaker (500 cm³)

☐ Test tube

☐ Heating tools

☐ Funnel

☐ Rubber plug

☐ Mortar

☐ Filter paper

☐ L-shaped glass tubing

☐ Rubber tubing

☐ Crucible

☐ Litmus paper

☐ Stand

☐ Sieve

☐ Evaporating dish

☐ Funnel and funnel stand

[Reagents]　☐ NaCl

☐ 3mol/L HCl

☐ Shell(or limestone)

☐ *conc.* H$_2$SO$_4$(or filling solution for car battery)

☐ *conc.* NH$_3$ water

☐ NaHCO$_3$(or baking powder)

☐ Plants ash

Procedure

I Acid HCl

1 Place 30 g of NaCl in a flat bottom flask, and attach a funnel pipe and an L-shaped glass tubing with a rubber plug.

2 Introduce *conc.* H_2SO_4 from the funnel, until the mass of NaCl is wholly immersed. The filling solution for car batteries is a good and useful substitute of reagent H_2SO_4. Then, heat the mixture slowly. Confirm the evolution of HCl gas with litmus paper (the color turns red) and/or *conc.* NH_3 water. White smoke of NH_4Cl evolves.

> WARNING and Additional Information : HCl gas must be treated with great care, for it is hygroscopic and IRRITATES THE TISSUE OF THE NOSE AND EYES. HCl gas evolves also from the reaction between ammonium chloride and H_2SO_4 as follows:
>
> $2NH_4Cl + H_2SO_4 \rightarrow 2(NH_4)_2SO_4 + 2HCl$
>
> NH_4Cl and $(NH_4)_2SO_4$ are used as nitrogen fertilizers.

3 Collect HCl gas evolved in a bottle by the "Downward-delivery" method (see p.10). HCl gas is heavier than the air and its vapor looks like a thin smoke in the air, because it changes to the mist of hydrochloric acid.

4 HCl gas dissolves well in water. If a sufficient amount of HCl gas is introduced into water, the resulting solution is concentrated hydrochloric acid. If an amount of HCl gas dissolved is not enough, it forms dilute hydrochloric acid

> Additional Information : Sulfuric acid is synthesized commercially from sulfur following the processes below:
>
> $S + O_2 \rightarrow SO_2$
>
> $2SO_2 + O_2 \rightarrow 2SO_3$
>
> $SO_3 + H_2O \rightarrow H_2SO_4$ (In practice, SO_3 is absorbed in *conc.* H_2SO_4 to produce *fuming* H_2SO_4.)
>
> Nitric acid is synthesized in large scale from ammonia following the processes below (called *Ostwald process*):
>
> $4NH_3 + 5O_2 \rightarrow 4NO + 6H_2O$
>
> $2NO + O_2 \rightarrow 2NO_2$
>
> $3NO_2 + H_2O \rightarrow 2HNO_3 + NO$

II   Alkalis
5   Synthesis of Ca(OH)$_2$
  (1) Place several pieces of shell(or limestone) in a mortar and grind
  to a powder.   Place this powder in a crucible and heat strongly for
  10 - 15 minutes.   The product is CaO.

  Additional Information : Shell(or limestone) mainly consists of
  calcium carbonate CaCO$_3$.   When CaCO$_3$ is heated to about 900 ℃,
  it decomposes into calcium oxide CaO and carbon dioxide as
  follows:
     CaCO$_3$ → CaO + CO$_2$

---

  (2) Add water to the product.   Pay attention that the reaction is
  strongly exothermic.   Add several drops of phenolphthalein solution
  to check that the reaction product is alkaline.   Separate the white
  precipitate Ca(OH)$_2$ from the solution.

6   Synthesis of NaHCO$_3$, Na$_2$CO$_3$ and NaOH
  (1) Place 100 cm^3 of *conc.* ammonia water in an Erlenmeyer flask
  and add about 40 g of NaCl till the solution becomes saturated.
  Introduce the upper part of the solution to another Erlenmeyer flask.
  Bubble CO$_2$ gas(evolved by the reaction of HCl and shell) to the
  solution.   Soon after the bubbling, no change in solution color or
  precipitation will be observed.   However, the reaction mixture gets
  gradually hotter and after about 30 minutes, a white precipitate,
  sodium hydrogencarbonate NaHCO$_3$ will be formed.
  (2) Place NaHCO$_3$(or baking powder) in an evaporating dish and
  heat it strongly.   The heated product is sodium carbonate Na$_2$CO$_3$.
  (3) Place 1 g of CaO in a beaker and add 10 cm^3 of water.   Then,
  add 1 g of Na$_2$CO$_3$ to the solution.   The white precipitate observed is
  calcium carbonate CaCO$_3$.
  (4) Filter out the precipitate.   The solution contains NaOH.   Keep
  the solution in a suitable plastic or glass bottle (but not in a PET
  bottle).

  Additional Information : Procedure 6(1) generating NaHCO$_3$ is
  called *ammonia soda process* (or *Solvay soda process*; Ernest
  Solvay (1838 - 1922) is a Belgian chemical engineer.).   NaHCO$_3$ is
  a main material for the glass industry.   In the present alkali
  industry, NaOH is synthesized by the electrolysis of NaCl dissolved
  in water.

---

7 Alkali from plant ash

(1) Remove bulky dust from plant ash with a sieve. Place about 50 g of plant ash thus refined in a 500 cm^3 beaker and pour 250 cm^3 of hot water and stir the mixture well. Take the upper transparent part of the solution by decantation.

(2) Filter the solution. Transfer the filtrate into an evaporating dish and heat to remove its moisture. The solid obtained is the crude potassium carbonate K_2CO_3. In most cases, the product is weakly colored. Recrystallize the crude product if necessary.

(3) Add HCl to the solid and confirm that CO_2 gas evolves.

· If coal ash is used in place of straw ash in the procedures above, what difference in reaction will be expected?

(4) Dissolve the solid into water and add $Ca(OH)_2$ solution.

· What is the precipitate formed?

Questions for students

1 Write the equation for the evolution of HCl in Procedure 2.

2 Write the equation for the reaction that CaO reacts exothermally with water in Procedure 5(2).

3 Write the equation for the reaction to produce $NaHCO_3$ in Procedure 6(1).

4 Write the equation for the reaction to produce NaOH in Procedure 6(3).

5 Write the equation for the reaction to produce KOH in Procedure 7(4).

Transition elements and their compounds
— Copper and silver

Purpose

Elements are classified into two categories; one is *representative elements* (*main group elements*) and the other *transition elements*. The former are elements in Groups 1, 2 and 12 - 18 of the *periodic table*, the latter those in Groups 3 - 11 of the periodic table. If one element of a group forms a compound with another element or ion, we can predict that the other members of the same group will show similar behavior. Copper and silver are very popular elements, and along with gold, they form Group 11 of the periodic table. Hence, their ions behave similarly. Confirm the similarity in the reactions of copper ion and silver ion.

Keywords

Transition element, Transition metal, Periodic table

Preparation

[Ware] □ Heating tools   □ Spatula
　　　 □ Test tube   □ Weighing paper
　　　 □ Beaker (100 cm^3)
　　　 □ Komagome pipette

[Reagents] □ $CuSO_4$ (copper (II) sulfate, anhydrous)
　　　 □ 0.1mol/L $AgNO_3$ (silver nitrate)
　　　 □ 1mol/L NH_3 (ammonia water)
　　　 □ 1mol/L NaOH (sodium hydroxide)
　　　 □ 0.1mol/L NaCl (sodium chloride)
　　　 □ 0.1mol/L KBr (potassium bromide)
　　　 □ 0.1mol/L KI (potassium iodide)
　　　 □ H_2S (hydrogen sulfide water; *i.e.* water saturated with H_2S, about
　　　　　0.1mol/L H_2S)

Additional Information : H_2S gas is prepared from the reaction between FeS and HCl or H_2SO_4.

$$FeS + 2HCl \rightarrow FeCl_2 + H_2S$$
$$FeS + H_2SO_4 \rightarrow FeSO_4 + H_2S$$

If generation of H_2S is difficult, then, dilute solutions of water-soluble sulfides such as $(NH_4)_2S$ or Na_2S are applicable.

Procedure

    I   Properties of Cu^{2+}

1   Place a spoonful of $CuSO_4$ (0.5 g) in a beaker and pour 50 cm³ of water to prepare solution of Cu^{2+} ion.

    · Compare the color of the powder of anhydrous copper sulfate and the color of the solution.

> WARNING : Be careful when handling H_2S gas and its solution, because H_2S is a DEADLY POISON.

2   Place 3 cm³ each of the above Cu^{2+} solution in three test tubes. Add drop by drop H_2S water, 1mol/L NH_3, and 1mol/L NaOH respectively to each test tube. Furthermore, add the same reagent in excess to each.

    · Describe the results.

        H_2S water _____

        1mol/L NH_3 _____

        1mol/L NaOH _____

3   Heat the mixture obtained by addition of 1mol/L NaOH.

    · Describe the results.   ☞ **See** Photograph ④

II   Properties of Ag⁺

4   Place 3 cm³ each of 0.1mol/L $AgNO_3$ in three test tubes.   Add drop
    by drop H_2S water, 1mol/L NH_3, and 1mol/L NaOH respectively to
    each test tube.   Furthermore, add the same reagent in excess to each.
    · Describe the results.

        H_2S water _____

        1mol/L NH_3 _____

        1mol/L NaOH _____

5   Heat the mixture obtained by addition of 1mol/L NaOH.
    · Describe the results.   ☞   **See**   Photograph ⑤

6   Place 3 cm³ each of 0.1mol/L $AgNO_3$ in three test tubes.   Add drop
    by drop 0.1mol/L NaCl, 0.1mol/L KBr, and 0.1mol/L KI respectively
    to each test tube.   Observe the change.

        NaCl _____

        KBr _____

        KI _____

    · Describe the color of each precipitate. ☞   **See**   Photograph ⑥

Questions for students

1 Why can you use aqueous solutions of $(NH_4)_2S$ or Na_2S instead of H_2S water? Is H_2S water acidic, neutral, or basic? How about the solutions of $(NH_4)_2S$ or Na_2S?

2 How can you explain the difference between the color of solid $CuSO_4$ (anhydrous) and its aqueous solution?

3 Write the equations for the following reactions for copper compounds.
  (1) The reaction between copper sulfate and hydrogen sulfide.

  (2) The reaction between copper sulfate and ammonia water.

  (3) The reaction between copper sulfate and sodium hydroxide.

  (4) The reaction between copper hydroxide and ammonia water.

  (5) The reaction of copper hydroxide by heating.

4 Write the equations for the following reactions for silver compounds.
  (1) The reaction between silver nitrate and hydrogen sulfide.

  (2) The reaction between silver nitrate and ammonia water.

  (3) The reaction between silver oxide and ammonia water.

  (4) The reaction of silver oxide by heating.

Separation and identification of cations

Purpose

A cation in the mixture can be separated by the reaction with an appropriate reagent as precipitate. This method is called *separation analysis* or *wet analysis*. Compare the reactions of dissolved cations with various reagents and apply this method to separate and identify several cations dissolved in water.

Keywords

Separation analysis, Wet analysis, Precipitate, Kipp's gas generator

Preparation

[Ware]　□ Heating tools　　　　　□ Beaker (100 cm^3)

　　　　□ Test tube　　　　　　　□ Glass stick

　　　　□ Spatula　　　　　　　　□ Komagome pipette

[Reagents]　□ *conc.* HCl (concentrated hydrochloric acid)

　　　□ 6mol/L HCl　　　　　　　　　□ 1mol/L HCl

　　　□ 6mol/L HNO$_3$ (nitric acid)　　　□ 1mol/L HNO$_3$

　　　□ 1mol/L H$_2$SO$_4$ (sulfuric acid)　　□ 6mol/L NH$_3$ (ammonia water)

　　　□ 1mol/L NaOH (sodium hydroxide)　□ 1mol/L NH$_3$

　　　□ 0.1mol/L AgNO$_3$ (silver nitrate)　□ 0.1mol/L Cu(NO$_3$)$_2$ (copper(II) nitrate)

　　　□ 0.1mol/L Pb(NO$_3$)$_2$ (lead(II) nitrate)　□ 0.1mol/L Fe(NO$_3$)$_3$ (iron(III) nitrate)

　　　□ 0.1mol/L FeSO$_4$ (iron(II) sulfate)　□ 0.1mol/L Zn(NO$_3$)$_2$ (zinc(II) nitrate)

　　　□ 0.1mol/L Al(NO$_3$)$_3$ (aluminum nitrate)　□ 0.1mol/L Ba(NO$_3$)$_2$ (barium nitrate)

　　　□ 0.1mol/L Ca(NO$_3$)$_2$ (calcium nitrate)　□ 0.1mol/L KSCN (potassium thiocyanate)

　　　□ 0.1mol/L K$_2$CrO$_4$ (potassium chromate)　□ H$_2$S (hydrogen sulfide)

　　　□ 0.1mol/L K$_4$[Fe(CN)$_6$] (potassium hexacyanoferrate(II))

　　　□ 0.1mol/L K$_3$[Fe(CN)$_6$] (potassium hexacyanoferrate(III))

　　　□ 0.1mol/L (NH$_4$)$_2$CO$_3$ (ammonium carbonate)

　　　□ Pb(CH$_3$COO)$_2$ paper (filter paper moistened by aqueous solution of lead(II) acetate)

　　　□ Sample solution (mixture of each 10 cm^3 of AgNO$_3$, Cu(NO$_3$)$_2$, Fe(NO$_3$)$_3$, and Zn(NO$_3$)$_2$ solutions listed above)

> Useful Information : For the purpose of carrying out only **Procedure** I described in the next page, metal chlorides except for AgCl are also usable instead of metal nitrates.

Procedure

I Comparison of the precipitation reactions of cations

1 Prepare nine test tubes, each containing 3 cm^3 of solutions of Ag$^+$, Cu^{2+}, Pb^{2+}, Fe^{2+}, Fe^{3+}, Zn^{2+}, Al^{3+}, Ba^{2+}, and Ca^{2+}, respectively. Add 1 cm^3 of 1mol/L HCl to each test tube.

2 Examine the reactions of nine kinds of ions with the solutions of H$_2$SO$_4$, HNO$_3$, NaOH, NH$_3$, and H$_2$S. For the reaction with H$_2$S, the change depends on whether the solution is acidic or basic. Compare the two results obtained in acidic and basic media. Also the change is compared for conditions whether NaOH or NH$_3$ is added in excess or not. The precaution concerning H$_2$S on p.44 should be taken.

3 Summarize the results as follows.

Reagents	Ag$^+$	Cu^{2+}	Pb^{2+}	Fe^{2+}	Fe^{3+}	Zn^{2+}	Al^{3+}	Ba^{2+}	Ca^{2+}
HCl									
H$_2$SO$_4$									
HNO$_3$									
NaOH									
NH$_3$									
H$_2$S									

NaOH, NH$_3$: UPPER/ small amount   H$_2$S : UPPER/ acidic
             LOWER/ excess   LOWER/ basic

II Separation and identification of cations from the mixture containing Ag$^+$, Cu^{2+}, Fe^{3+}, and Zn^{2+} (see **Flow sheet of analysis** on p.50)

4 Place 40 cm^3 of sample solution (see [Reagents] on p.47) in a beaker and add 2 cm^3 of 6mol/L HCl. Mix and wait for a while so that the reaction is complete. The precipitate (called Precipitate 1) is filtered off.

48

5   Prepare H2S gas from the reaction between FeS and HCl (or H2SO4) by means of Kipp's gas generator.   Bubble H2S into the filtrate and the precipitate formed by bubbling (called Precipitate 2) is filtered off.

6   Boil the filtrate and expel the dissolved H2S so as the lead acetate test paper does not turn black.   Next, add several drops of 6mol/L HNO3 and add 10 cm³ of 6mol/L NH3.   The precipitate (called Precipitate 3) is filtered off.

7   Add H2S solution to the filtrate to make Precipitate 4.

8   Identification of Precipitate 1 :   Place Precipitate 1 in a test tube. Add a small amount of 6mol/L NH3 to dissolve the precipitate. Next, add a small amount of 6mol/L HNO3 to this solution and observe the change.

9   Identification of Precipitate 2 :   Place Precipitate 2 in a beaker. Add 5 cm³ of 6mol/L HNO3 and heat to dissolve the precipitate.   Next, cool the solution and add 10 cm³ of 6mol/L NH3 and observe the change of color.

10   Identification of Precipitate 3 :   Place Precipitate 3 in a test tube. Add 4 cm³ of 6mol/L HCl to dissolve the precipitate.   Divide the solution to two test tubes.   Add several drops of $K_4[Fe(CN)_6]$ solution to one and observe.   Add several drops of KSCN solution to the other and observe.

Caution :   Do not add too much 6mol/L HCl in Procedure 4.   If the precipitate is colloidal, heat the solution to facilitate the filtration. The particles in a colloidal state are coagulated by heating.

Questions for students

1   Fill in the blanks in **Flow sheet of analysis** on next page.

2   Explain the reason why nitric acid is added in Procedure 6.

3   Write the equations for the following reactions in Procedure 8.
   (1) The dissolution of Precipitate 1.

   (2) The precipitation by adding nitric acid.

4   Write the ionic equation for the dissolution of Precipitate 2 by adding an excess amount of ammonia water in Procedure 9.

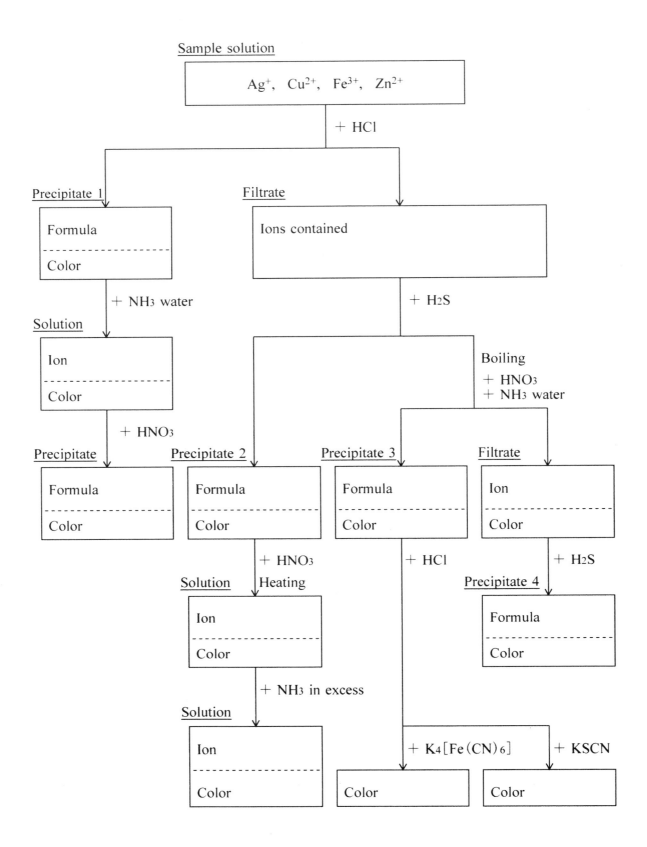

Flow sheet of analysis

Alcohols, aldehydes, carboxylic acids, and esters

Purpose

Organic substances are usually composed of only a few elements but their number is indefinitely large. They are formed mainly by covalent bonds. Properties of organic compounds are significantly different from those of inorganic compounds and functional groups play key roles in determining their physicochemical properties. Alcohol molecules contain hydroxy group, $-OH$, and aldehydes contain carbonyl group, $-C=O$, while carboxylic acids and esters are characterized by carboxyl group, $-COOH$, and ester group, $-COOR$, respectively. Examine the properties of alcohols and aldehydes in Procedures I and II, and also the properties of acetic acid as the representative of carboxylic acids in Procedure III. Synthesize ethyl acetate from the reaction between acetic acid and ethanol in Procedure IV.

Keywords

Alcohol, Aldehyde, Carboxylic acid, Ester, Functional group,
Esterification, Ammoniac silver nitrate solution, Silver mirror reaction

Preparation

[Ware]　□ Heating tools
　□ Test tube
　□ Evaporating dish
　□ All-purpose pH test paper
　□ Beaker (50, 200 cm³)
　□ Graduate (10 cm³)

　□ Thermometer (0 - 100 ℃)
　□ Spatula
　□ Komagome pipette
　□ Forceps
　□ Glass stick
　□ Boiling stone

[Reagents]　□ CH_3OH (methanol)
　□ C_2H_5OH (ethanol)
　□ C_4H_9OH (1-butanol)
　□ $C_3H_5(OH)_3$ (glycerin)
　□ 3mol/L H_2SO_4
　□ $CH_3COOC_2H_5$ (ethyl acetate)

　□ 1mol/L NH_3 (ammonia water)
　□ 0.1mol/L $AgNO_3$ (silver nitrate)
　□ Na (sodium)
　□ conc. H_2SO_4 (concentrated sulfuric acid)
　□ 2mol/L $NaOH$ (sodium hydroxide)
　□ Na_2CO_3 (sodium carbonate, anhydrous)

　□ Cu (copper wire, winded into spiral form about 2 cm in length)
　□ $HCHO$ (formalin; 37% formaldehyde solution)
　□ CH_3COOH (acetic acid, glacial; see **Appendix** on p.63)
　□ CH_3COONa (sodium acetate)

51

Procedure

    I   Properties of alcohols

1   Place about each 1 cm^3 of ethanol, 1-butanol, and glycerin in three separate test tubes. Observe the color, smell, acidity and solubility to water.

Alcohols	Color	Smell	Acidic/Neutral/Basic	Solubility to water
C_2H_5OH	_____	_____	_____	_____
C_4H_9OH	_____	_____	_____	_____
$C_3H_5(OH)_3$	_____	_____	_____	_____

2   Place about 1 cm^3 of methanol and ethanol in two separate evaporating dishes and ignite them both.

    · How do methanol and ethanol burn?

3   Place about 3 cm^3 of ethanol in a test tube and add a rice-grain size of sodium using forceps.

    · Describe the reaction.

> WARNING : When using volatile alcohols, EXTINGUISH ALL FLAMES. Handle sodium carefully because it is HIGHLY REACTIVE.

    II   Formation and properties of formaldehyde

4   Place about 1 cm^3 of methanol in a test tube. Shake it well until the interior wall of the test tube is very wet.

5   Heat a winded copper wire until it glows. Then, leave it in the air for a while so that the surface of the copper wire becomes darkish. Introduce the copper wire into the test tube without dipping it into the methanol. Then take out the wire. Repeat introducing and taking out the wire several times. Do the steps mentioned above a few times and check the smell of the product.

    · What change takes place in the copper wire?

    · What is the smell of the product?

    · Write the chemical formula and the name of the product.

6　Place 10 cm³ of 0.1mol/L AgNO₃ in a test tube and add 1mol/L NH₃ drop by drop.　At first precipitate is formed.　Continue adding NH₃ water until the formed precipitate dissolves completely.　What you have prepared is *ammoniac silver nitrate solution.*

7　Add 2 cm³ of ammoniac silver nitrate solution to the test tube treated in Procedure 5.　Warm it lightly in a hot water bath.

☞ **See** Photograph ⑦

・Account for the results you have observed.

8　Place several drops of methanol and formalin separately in two test tubes.　Add 2 cm³ of ammoniac silver nitrate solution to each of the test tubes and warm them in a hot water bath.　Compare the result with Procedure 7.

・Account for the results you have observed.

> Caution :　Do not store the leftover ammoniac silver nitrate solution, for it may generate explosive compounds.　The silver that adheres to the interior wall of the test tube can be dissolved in a small amount of *conc.* nitric acid.

III　Properties of acetic acid and its strength as an acid

9　Place about 1 cm³ of glacial acetic acid in a 50 cm³ beaker and check the color and the smell.　Add about 10 cm³ of water and check the acidity of the solution with a piece of all-purpose pH test paper.

> WARNING :　Avoid breathing the vapor of glacial acetic acid and skin contact.　Glacial acetic acid ATTACKS THE SKIN.　WASH IT AWAY WITH WATER IMMEDIATELY if it comes into contact with any part of your body.

・What is the color and the smell of acetic acid?

・What is the acidity of acetic acid?

10　Add about 0.2 g of Na₂CO₃ to acetic acid solution little by little.

・Describe the reaction.

11　Place about 2 g of CH₃COONa in a 50 cm³ beaker and add about 5 cm³ of 3mol/L H₂SO₄.　Check the smell.　Warm the mixture in warm water if there is no smell.

・How does the smell change by adding dilute sulfuric acid?

IV   Synthesis of ethyl acetate

12   Place 2 cm^3 of C_2H_5OH in a dry test tube and add 2 cm^3 of glacial acetic acid.   While cooling this test tube with water, gently add 0.5 cm^3 of *conc.* H_2SO_4 by letting it flow down the surface of the glass stick.   In adding *conc.* H_2SO_4, properly tilt the test tube.

13   Add 2 or 3 pieces of boiling stone to the test tube and heat the mixture in hot water at 70 - 80 ℃ for about 5 minutes.

14   Add about 10 cm^3 of water to the test tube after heating.   The product will be separated into the upper layer.

・What is the color of ethyl acetate?   What does it smell like?

Questions for students

1   Why are the alcohols formed by a small number of carbon atoms water-soluble?   Which flame is more colorless, methanol or ethanol?   Account for the reasons.

2   Write the equation for the reaction between ethanol and sodium.

3   In Procedure 5, what effect does copper have on methanol?

4   What is the name of the reaction in Procedure 7?   What is the functional group that causes this reaction?

5   Write the equations for the following reactions.
　(1) Ionic equation for the electrolytic dissociation of acetic acid.

　(2) The reaction between sodium carbonate and acetic acid.

　(3) The reaction between sodium acetate and sulfuric acid.

　(4) The production of ethyl acetate from acetic acid and ethanol.

6   From the results of Procedures 10 and 11, arrange carbonic acid, sulfuric acid, and acetic acid in order of the acid strength.

7   As for the synthesis of ester like ethyl acetate, what kind of role does sulfuric acid play?

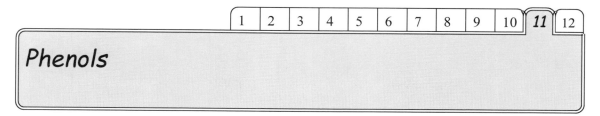

Phenols

Purpose

Aromatic compounds are generally derived from benzene. They are important to many chemical industries such as rubber, plastics, fibers, explosives, paint, and petroleum. Substituted aromatic compounds are characterized by the functional group. When hydroxy groups are combined to the benzene ring, the resulting compounds will be slightly acidic. The simplest case is phenol. Since properties of aromatic hydroxy compounds differ somewhat from those of most alcohols, they are classed as phenols. Explore the properties of phenols and synthesize methyl salicylate from salicylic acid.

Keywords

Aromatic compound, Benzene ring, Phenols

Preparation

[Ware]　□ Test tube

　　□ Beaker（100, 200 cm^3）

　　□ Komagome pipette

　　□ Thermometer

[Reagents]　□ *conc.* H_2SO_4（concentrated sulfuric acid）

　　□ 1mol/L NaOH（sodium hydroxide）

　　□ 2mol/L HCl（hydrochloric acid）

　　□ 2mol/L CH_3COOH（acetic acid）

　　□ 0.1mol/L $FeCl_3$（iron（Ⅲ） chloride）

　　□ $NaHCO_3$（saturated solution of sodium hydrogencarbonate）

　　□ Br_2（bromine water）

　　□ C_6H_5OH（phenol）

　　□ 1% C_6H_5OH（aqueous solution of phenol）

　　□ $C_6H_4(CH_3)(OH)$（cresol, any one of the isomers （*o*-, *m*-, and *p*-cresol） or the mixture）

　　□ $C_6H_4(OH)COOH$（salicylic acid）

　　□ CH_3OH（methanol）

Procedure

  I   Properties of phenols

┌───┐
  WARNING : PHENOL ATTACKS THE SKIN AND CLOTHES.
└───┘

1 Leave a bottle of phenol in hot water to melt. The melting point of phenol is 41 °C, and the density 1.07 g/cm³. Place about 1 cm³ of phenol in a test tube and check the color and smell.
   · What is the color and smell of phenol? ☞ **See** Photograph ⑧

2 Add about 5 cm³ of water to phenol and warm the mixture in hot water of 70 - 80 °C. Cool the resulting homogeneous solution in water and observe the change.
   · Compare the solubilities of phenol in both hot water and cold water.

3 Place 5 cm³ of aqueous solution of phenol in another test tube. Add 1 cm³ of 1mol/L NaOH to it and mix well. Next, add about 1 cm³ of 2mol/L CH_3COOH and observe the change.
   · Describe the change induced by adding 1mol/L NaOH.

   · Describe the change induced by further adding 2mol/L CH_3COOH.

4 Add about 5 cm³ of water to the remaining solution (1 cm³) of phenol and mix. Next, add several drops of 0.1mol/L $FeCl_3$ to the reaction mixture and observe the change.

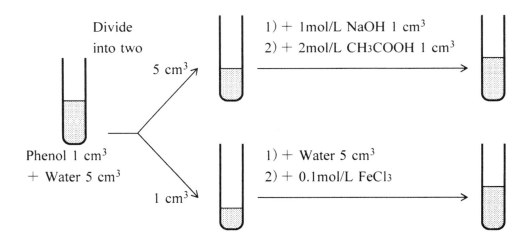

5 Prepare two test tubes. Place several drops of cresol in one test tube and a spoonful of salicylic acid (about 0.1 g) in the other. Add about 5 cm^3 of water to each and mix them well. Next, add several drops of 0.1mol/L FeCl$_3$ to them and observe the change of color.

・Compare the change in color for cresol and salicylic acid.

6 Place about 2 cm^3 of 1% aqueous solution of phenol in a test tube. Add about 1 cm^3 of bromine water to it little by little and shake. Observe the change.

WARNING : BROMINE ATTACKS THE SKIN AND THROAT.

II   Synthesis of methyl salicylate

7 Place 0.5 g of salicylic acid and 3 cm^3 of methanol in a dry test tube. Add about 0.5 cm^3 of *conc.* sulfuric acid to it. Warm the reactant for about 5 minutes in hot water of 50 - 60 ℃ while shaking well. Place 30 cm^3 of saturated solution of NaHCO$_3$ in a beaker. Pour the reaction mixture into the beaker little by little. The oily product will be formed in the bottom of the beaker. Check the smell of the product.

・What is the smell of the product?

Questions for students

1　Write the chemical formula and the name of the product of the reaction between phenol and sodium hydroxide.

2　Using structural formula write the equations for the following reactions.

（1）　The reaction between sodium phenoxide dissolved in water and acetic acid.

（2）　The reaction for the production of 2,4,6-tribromophenol from phenol and bromine.

（3）　The reaction for the production of methyl salicylate from salicylic acid and methanol.

3　Account for the reason why the reaction mixture is poured into a saturated solution of sodium hydrogencarbonate in Procedure 7.

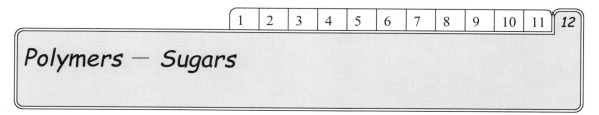

Polymers — Sugars

Purpose

The simplest sugars, called monosaccharides, are polyhydroxy aldehydes or ketones and they are usually formed of five or six carbon atoms. Glucose and sucrose are the representatives of sugars. The former occurs naturally in fruits, plants, honey, and in the blood and urine of animals, while the latter is common table sugar. Starch is a polysaccharide, which is stored in the seeds, roots, and fibers of plants. Compare the reduction abilities of the three kinds of sugar and also their reactions with iodine.

Keywords

Sugar, Starch-iodine reaction, Reducing sugar

Preparation

[Ware] □ Heating tools

　□ Test tube

　□ Beaker (100 cm^3)

　□ Komagome pipette

[Reagents] □ CuSO$_4$·5H$_2$O (copper (II) sulfate pentahydrate)

　□ KNaC$_4$H$_4$O$_6$·4H$_2$O (potassium sodium tartrate tetrahydrate, *Rochelle salt*)

　□ NaOH (sodium hydroxide)

　□ 1% Glucose solution

　□ 1% Sucrose solution

　□ 1% Starch solution

　□ 3mol/L H$_2$SO$_4$ (sulfuric acid)

　□ Na$_2$CO$_3$ (sodium carbonate anhydrous, powder)

　□ Iodine and potassium iodide solution (For preparation, see **Appendix** on p.63.)

Procedure

I   Reduction and hydrolysis of sugars

1   Prepare Fehling's solution as follows:

Solution A: Add 6.9 g of $CuSO_4 \cdot 5H_2O$ to water less than 100 cm^3 to make a 100 cm^3 solution.

Solution B: Add 35 g of $KNaC_4H_4O_6 \cdot 4H_2O$ and 10 g of NaOH to water less than 100 cm^3 to make a 100 cm^3 solution.

Place 5 cm^3 of both solution A and solution B in a test tube and mix them.   This is *Fehling's solution.*

─ Fehling's solution : Fehling's solution, which was invented by Hermann von Fehling (1812 - 1885), is one of the testing reagents for reducing sugars.   The reaction process involves the oxidizing action of complex copper (Ⅱ) ion; a red precipitate Cu_2O (copper (Ⅰ) oxide) indicates the oxidation of an aldehyde or other easily oxidizable substances.   ☞ **See** Photograph ⑨

2   Place 3 cm^3 each of 1% glucose solution, 1% sucrose solution, and 1% starch solution in three separate test tubes.   Add 1 cm^3 of Fehling's solution to each test tube.   Heat them and observe the changes.   If the change is not apparent, then heat again for a few more minutes.

3   In (a) different test tube/tubes, place 3 cm^3 of solution/solutions for which the change was observed in Procedure 2.   Add 1 cm^3 of 3mol/L H_2SO_4 to this/these and heat it/them gently without boiling for about 5 minutes.   Summarize the observations in Procedures 2 and 3 as follows.

	Glucose solution	Sucrose solution	Starch solution
Procedure 2	_____	_____	_____
Procedure 3	_____	_____	_____

4   Add Na_2CO_3 powder little by little to the mixture until generation of CO_2 bubbles ceases.   Add about 1 cm^3 of Fehling's solution to the reaction mixture.   Heat it and observe the change.

II Starch-iodine reaction and hydrolysis of sugars

5 Place 1 cm³ of 1% glucose solution, 1% sucrose solution, and 1% starch solution in three separate test tubes. Add 2 - 3 drops of *iodine and potassium iodide solution* to each.

  — Iodine and potassium iodide solution : This solution is used as an indicator for polysaccharides and oligosaccharides, and turns blue by adding starch, reddish violet by dextrin, and reddish brown by glycogen.   ☞ **See** Photograph ⑩

6 Place 1 cm³ of sample solution/solutions for which the change(s) was/were observed in Procedure 5 in another test tube(s). Add 2 - 3 drops of iodine and potassium iodide solution and 1 cm³ of 3mol/L H_2SO_4 to this/these. Heat the mixed solution/solutions without boiling. Observe the change.

III Starch-iodine reaction

7 Place 1 cm³ of 1% starch solution in a test tube and add 2 - 3 drops of iodine and potassium iodide solution.

8 Heat the mixture gently without boiling. Next, cool it with water and observe the change.

Questions for students

1 Write the chemical formula of the red precipitate formed by the reaction between reducing sugars and Fehling's solution.

2 Why do the solutions of sucrose and starch show the ability to reduce other compounds when heated with dilute sulfuric acid?

3 Account for the reason why the changes were observed in Procedure 6.

4 Starch-iodine reaction occurs when the iodine molecule is taken in the hollow part of the spiral of starch molecule. Explain the color change in the course of heating and cooling of the reactant in Procedure 8.

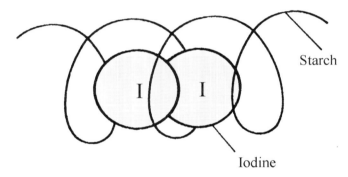

Appendix

1 Reagent Solutions Frequently Used in Laboratory and Their Preparation

Acids :

conc. HCl :   Commercially available.   12mol/L (12M), density 1.19 g/cm^3

◇ 6mol/L HCl :   Dilute 1 volume of *conc.* HCl with 1 volume of water.

conc. H$_2$SO$_4$:   Commercially available. (96 - 98%)   18mol/L (18M), density 1.84 g/cm^3

◇ 3mol/L H$_2$SO$_4$:   Dilute 1 volume of *conc.* H$_2$SO$_4$ with 5 volumes of water.

> WARNING :   Pour *conc.* H$_2$SO$_4$ into water.   NEVER POUR WATER into *conc.* H$_2$SO$_4$!
> Dilution of H$_2$SO$_4$ is HIGHLY EXOTHERMIC.

conc. HNO$_3$:   Commercially available.   16mol/L (16M), density 1.42 g/cm^3

◇ 6mol/L HNO$_3$:   Dilute 38 cm^3 of *conc.* HNO$_3$ with water to 100 cm^3.

glacial CH$_3$COOH :   Commercially available.   17mol/L (17M);   Aqueous solution of acetic acid with purity over 96%.   The freezing point of pure acetic acid is 17 ℃.   The adjective *glacial* refers to the fact that acetic acid with high purity kept in the laboratory freezes in winter.

◇ 6mol/L CH$_3$COOH :   Dilute 35 cm^3 of glacial acetic acid with water to 100 cm^3.

Alkalis :

conc. NH$_3$ water :   Commercially available.   15mol/L (15M), 28%, density 0.90 g/cm^3

◇ 6mol/L NH$_3$:   Dilute 1 volume of *conc.* NH$_3$ water with 1.5 volumes of water.

NaOH, KOH :   HIGHLY HYGROSCOPIC.   Weigh quickly.

◇ 6mol/L NaOH :   Dissolve 24 g of NaOH to water and make 100 cm^3 of solution.

◇ 6mol/L KOH :   Dissolve 34 g of KOH to water and make 100 cm^3 of solution.

Lime water :   Add an excess amount of Ca(OH)$_2$ to water and take the supernatant part.

Indicators and Other :

Phenolphthalein (PP) solution :   Dissolve 1 g of PP to 90 cm^3 of 95% ethanol, then add water to 100 cm^3.

Bromothymol Blue (BTB) solution :   Dissolve 0.1 g of BTB to 20 cm^3 of 95% ethanol, then add water to 100 cm^3.

3% H$_2$O$_2$ solution :   Dilute 1 volume of 30% H$_2$O$_2$ solution (commercially available) with 9 volumes of water.

Iodine and potassium iodide solution :   Dissolve 1.5 g of KI in 100 cm^3 of water, and add 0.3 g of I$_2$ to the solution.

2 SI Base Units, SI Derived Units, and SI Prefixes

SI Base Units

Quantity	Symbol
Length	m (meter)
Mass	kg (kilogram)
Time	s (second)
Current	A (Ampère)
Temperature	K (Kelvin)
Amount of substance	mol (mole)

SI Derived Units

Quantity	Derived units	Representation in base units	
Force	N (Newton)	$m \cdot kg \cdot s^{-2}$	
Pressure	Pa (Pascal)	$N \cdot m^{-2}$	$= m^{-1} \cdot kg \cdot s^{-2}$
Energy	J (Joule)	$N \cdot m$	$= m^2 \cdot kg \cdot s^{-2}$
Power	W (Watt)	$J \cdot s^{-1}$	$= m^2 \cdot kg \cdot s^{-3}$
Charge	C (Coulomb)	$A \cdot s$	
Voltage	V (Volt)	$J \cdot C^{-1}$	$= m^2 \cdot kg \cdot s^{-3} \cdot A^{-1}$

SI Prefixes

Symbol	Power	Symbol	Power
da (deca)	10^1	d (deci)	10^{-1}
h (hecto)	10^2	c (centi)	10^{-2}
k (kilo)	10^3	m (mili)	10^{-3}
M (Mega)	10^6	μ (micro)	10^{-6}
G (Giga)	10^9	n (nano)	10^{-9}
T (Tera)	10^{12}	p (pico)	10^{-12}
P (Peta)	10^{15}	f (femto)	10^{-15}

3 Physical Constants and Their Typical Symbols

Speed of light in vacuum, c : 3.00×10^8 m/s

Avogadro constant, N_A : 6.02×10^{23} /mol

Elementary charge, e : 1.60×10^{-19} C

Mass of electron, m_e : 9.11×10^{-31} kg

Mass of proton, m_p : 1.67×10^{-27} kg

Mass of neutron, m_n : 1.67×10^{-27} kg

Gas constant, R : 8.31 J/(mol·K) $= 8.31 \times 10^3$ Pa·L/(mol·K)

$(\because$ Pa $=$ N/m^2, J $=$ N·m, 1 m^3 $= 10^3$ L$)$

Molar volume of ideal gas   22.4 L (dm^3)/mol at 273.15 K, 101325 Pa

4 Index and Logarithm

Index: $10^n = 10 \times 10 \times 10 \times ... \times 10$ (The numeral 10 is repeated n times.)

$10^{-n} = 1/10^n$

10^n : ten to the nth power (the nth power of ten), ten to the power of n

10^{-n} : ten to the minus (negative) nth power, the minus (negative) nth power of ten

Similarly, $a^n = a \times a \times a \times ... \times a$ (a is repeated n times.) $a^{-n} = 1/a^n$

$10^m \times 10^n = 10^{m+n}$ and $10^m/10^n = 10^{m-n}$

Example: $2.5 \times 10^{-2} = 2.5 \times (1/10^2) = 2.5/100 = 0.025$

Logarithm: $10^n = N \iff n = \log_{10}N$ Therefore, $\log_{10}10^{-2} = -2$

$\log_{10}N$: logarithm of N (to the base ten)

5 Significant Figures and Errors

Accuracy and precision : The exactness of measurements is an important part in experiments and is indicated by the number of significant figures. The examples of significant figures are as follows: 12 : two significant figures, 34.5 : three significant figures,

6.700 : four significant figures, 0.0089 : two significant figures

Measured values will contain errors. For example, if a value 20.5 is obtained, the true value x is in the range of $20.5-0.05 \leqq x < 20.5+0.05$. If x is 20.50, then $20.50-0.005 \leqq x < 20.50+0.005$. Thus the sum $(20.5+0.24)$ should be calculated as 20.7.

$\because$ $x=20.5$, $y=0.24$ $\Rightarrow$ $(20.5-0.05)+(0.24-0.005) \leqq x+y < (20.5+0.05)+(0.24+0.005)$

$\therefore$ $20.685 \leqq x+y < 20.795$

In this sum, the whole number part is certain, while the tenths place is a little uncertain and the hundredths place is meaningless. The figures are, therefore, significant to the tenths place, 20.7. Also, the product of the two measured values is taken as 4.9 instead of 4.92.

```
  2 0.[5]              2 0.[5]
+)  0.2[4]           ×)  0.2[4]
  2 0.[7]{4}            [8]{2 0}
∴ 2 0. 7             4 [1][0]        [ ] : uncertain figure
                     4.[9]{2 0}  ∴ 4. 9   {~} : meaningless figure
```

Arithmetic calculation : For addition and subtraction, the result should have as many decimal places as the measured number with the smallest number of decimal places. For multiplication and division, the result should have as many significant figures as the measured number with the smallest number of significant figures.

Scientific notation : In chemistry we encounter very large numbers, such as the Avogadro constant and very small numbers such as mass of an atom. Scientific notation ($a \times 10^b$) is a convenient system of expressing very large or very small numbers. In this notation, only significant figures are shown, for example, 60200 is expressed as 6.02×10^4.

6 Graphical Expression

Very often, experimental results are described more clearly by graphs. An example of graphical expression is given below.

Remarks in use of graph are as follows:
1) Data are plotted with appropriate keys such as ●, ○, ×, or △. The size of them corresponds to the precision of measurements.
2) Scales are assigned to *x* and *y* axes. Titles of *x* and *y* axes are expressed with units used.
3) Legend (caption) is arranged in the lower side of the graph.

Table 1   Variation of Water Density with Temperature

Temperature[°C]	0	2	4	6	8	10	12	14	16	18	20
Density [g/cm³]	0.99981	0.99991	0.99997	0.99994	0.99985	0.99970	0.99949	0.99924	0.99894	0.99860	0.99820

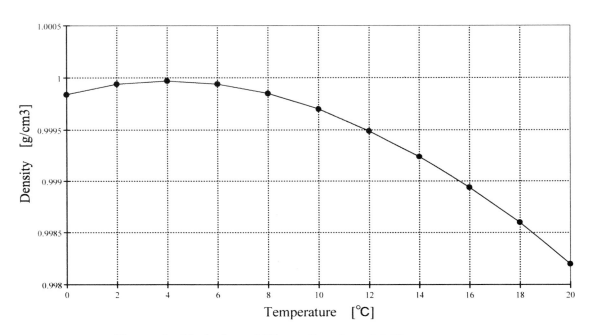

Figure 1   Variation of Water Density with Temperature

7 Exercises to Check Your Understanding of Chemistry and English

— May you improve your ability of scientific English!

1 Fill in each blank with the correct word to complete the sentences.

(1) Nearly [   ] quarters of the Earth's crust is made of oxygen and [   ].

(2) A chemical element is identified by the atomic number that is the number of [   ] in the atoms of the element. Despite the change of properties of a substance by reaction, elements in the substance are [   ]. (保存される) (➡ p.13)

(3) A molecule of water [   ] (…でできている) of one atom of oxygen and two atoms of hydrogen.

(4) One part by weight of hydrogen combines with eight parts by weight of oxygen to form water. Then, two parts by [   ] of hydrogen combine with [   ] part by [   ] of oxygen to form water.

(5) Organic compounds are [   ] (…でできている) up of carbon, hydrogen, oxygen, nitrogen, and some other atoms. (➡ p.51)

(6) The elements in Group 17 of the periodic table are called the [   ]. (➡ p.43)

(7) Sodium is a metal, and chlorine is a [   ]. They react together to form sodium chloride, that is a typical [   ] compound. (➡ p.43)

(8) Sodium chloride is also formed by neutralization of HCl with NaOH. Neutralization is defined as the reaction between [   ] ion and [   ] ion to produce water. (➡ p.35)

(9) [   ] is defined as the loss of electrons by an atom, molecule, or ion, and [   ] as the gain of electrons.

(10) A dissolved substance raises the [   ] point and lowers the [   ] point of a solvent.

2 Fill in each blank with the name of chemist.

(1) Ammonia is synthesized industrially in large quantities by the [   ] process.

(2) The quantity of electric charge that one mole of electron has is called the [   ] constant.

(3) It was not by mere intuition that [   ] proposed for benzene the formula of cyclohexatriene.

3 Choose the correct one in the parenthesis.

(1) Each test tube and each beaker (was / were) rinsed with pure water.

(2) Five pellets of NaOH granules (was / were) added.

(3) Melting point and boiling point (was / were) inspected.

4 Fill in each blank with the correct word to spell out the number or formula.

(1) 6.02×10^{23} ⇒ Six point zero two [＿＿＿＿] ten to the twenty-third [＿＿＿＿]

⇒ Six point zero two [＿＿＿＿] by ten to the twenty-third [＿＿＿＿]

(2) x/y ⇒ x [＿＿＿＿] by y

⇒ x [＿＿＿＿] y

(3) $x^2 - y^2 = (x + y)(x - y)$

⇒ x squared minus y squared equals brackets x plus y [＿＿＿＿] brackets x minus y

⇒ x squared minus y squared equals x plus y in brackets [＿＿＿＿] x minus y in brackets

⇒ x squared minus y squared equals the [＿＿＿＿] of x plus y, and x minus y

5 Read the information given below. By considering the difference in meaning of the underlined parts in it, choose (a) suitable sentence(s) out of the descriptions (1) to (4).

Phenol and bromine <u>will react</u> to produce 2,4,6-tribromophenol. On the contrary benzene and hydrogen <u>could react</u> to produce cyclohexane. (➡ p.57)

(1) When bromine water is poured into aqueous solution of phenol, 2,4,6-tribromophenol is observed spontaneously as the product.

(2) When hydrogen is bubbled into benzene, cyclohexane is observed spontaneously as the product.

(3) In order to transform benzene to cyclohexane, benzene and hydrogen must be treated in the presence of Ni catalyst at high temperatures and pressures.

(4) As to the reactions of aromatic compounds, addition reactions to benzene ring occur more easily than substitution reactions.

Answer

1 (1) three, silicon   (2) protons, conserved   (3) consists   (4) volume, one, volume

(5) made   (6) halogens   (7) non-metal, ionic   (8) hydrogen, hydroxide

(9) Oxidation, reduction   (10) boiling, freezing

2 (1) Haber   (2) Faraday   (3) Kekulé

3 (1) was   (2) were   (3) were

4 (1) times, power, multiplied, power   (2) divided, over   (3) times, times, product

5 (1), (3)

Chemistry Laboratory
for Secondary and Higher Education
3rd Edition

2004 年 4 月 10 日	第 1 版	第 1 刷	発行
2006 年 3 月 30 日	第 1 版	第 2 刷	発行
2010 年 4 月 10 日	第 2 版	第 1 刷	発行
2013 年 3 月 30 日	第 2 版	第 2 刷	発行
2020 年 3 月 20 日	第 3 版	第 1 刷	印刷
2020 年 3 月 30 日	第 3 版	第 1 刷	発行

著　　者　　園 部 利 彦
　　　　　　川 泉 文 男

発 行 者　　発 田 和 子

発 行 所　　株式会社　学術図書出版社

〒113−0033　　東京都文京区本郷 5 丁目 4 の 6
TEL 03−3811−0889 振替　00110−4−28454
印刷　サンエイプレス (有)

定価は表紙に表示してあります.

©2004,2010,2020　　T. SONOBE,　F. KAWAIZUMI　　Printed
in Japan
ISBN978−4−7806−0848−9　　C3043